W9-BYL-027

The Patent Files

ALSO BY DAVID LINDSAY

Madness in the Making:
The Triumphant Rise and Untimely
Fall of America's Show Inventors

The Patent Files

DISPATCHES FROM THE FRONTIERS OF INVENTION

DAVID LINDSAY

THE LYONS PRESS

The Lyons Press
123 West 18 Street
New York, New York 10011

Design by Cindy LaBreacht
Printed in the United States of America

10 9 8 7 6 5 4 3 2 1

Library of Congress Cataloging-in-Publication Data

Lindsay, David, 1957–
The patent files: dispatches from the frontiers of invention / David Lindsay.
p. cm.
ISBN 1-55821-741-X
1. Inventions—United States—Anecdotes. 2. Inventors—United States—Anecdotes. I. Title.
T21.L463 1999
609.73—dc21 98-38808
CIP

To Leslie
For the inspiration
The perspiration
Everything

Contents

Foreword

THE MOLDY OLD CHESTNUT has it that necessity mothers invention. But David Lindsay's "The Patent Files" came to *New York Press* in the fall of 1993 by the grace of that noun's far more joyous cousin, felicity. An alternative weekly newspaper in New York City, after all, doesn't *need* a column devoted to the culture and business of the United States Patent Office, no more than it does one written by a dominatrix or a househusband in suburban New Jersey. No right-thinking editor would sit about the office and conjure such a thing, then go out into the wilds to find a writer, force the idea upon him and believe that the commissioned column will shine. No, a writer must come to this sort of idea himself, or come close to it, then find the editor who can tease it out of his or will, in any event, let him write it.

Thus it was with David and *Press*'s editor and owner, Russ Smith, late that early-1990s summer. This book comprises the work David did in the subsequent four and a half years, time spent surveying a world of patents that, as he'll be the first to tell you, is based on confusion and disarray.

Or at least *seems* to be based on confusion and disarray.

"At first glance," wrote David in one of his earliest columns, "the mass of issued patents reads like chaos. Somebody seems to be blowing free jazz improv into your already scattershot day, jamming up psychic space with fragments of metal and mesh. But gradually, patterns begin to take shape. Little obsessions pop up and refuse to go away, like McGuffins in the tracts of an uncertain progress."

These patterns, and the obsessions that David held up to scrutiny, are "The Patent Files." As one of the earliest columns puts it, patents for David make up a sort of encyclopedia of the collective unconscious; they embrace "quirks, poems, late-night jokes, brilliant flourishes and shimmering nightmares. Trash bags for use in outer space, coffin alarms for the prematurely buried, the simple light bulb, recombinant DNA—it's all on the record."

Most of it, anyway. Here's some that's been missing heretofore: David had written for *Press* before "The Patent Files," first in 1990 as a sort of accidental fiction writer who'd mistakenly sent the wrong short story to the paper's then associate editor, Phyllis Orrick, and then for a short time as a restaurant critic. Previously, he'd been a musician, playing bass in bands for about ten years in Wisconsin, Boston and New York City (where the Klezmatics recorded his "Man in a Hat" and "Ode to Karl"), and an ersatz art critic for a samizdat 'zine put out by They Might Be Giants. The fake art criticism, David says, was meant to read like a French deconstructivist who was a little too deep into his brandy—and by all accounts it did. On the strength of it, he got some work with *Ear Magazine*, writing for no pay. And then, after a few years' work as managing editor for a couple of nearly fictitious magazines (trade publications made up of phony circulation numbers and mind-numbing product-driven prose), he entered the world of freelance journalism. The *Press* story followed, as well as pieces in the *Wall Street Journal* and the *Village Voice*.

Also, inventions. David has no background in science: in college, he received a D in geology and was once told by a physics teacher that he'd been watching too much *Star Trek*, but he'd always been interested in how things—songs, arguments, religions, parties—were put together. His obsession with inventions came from this, and eventually he attempted a few of his own.

One of these midnight philosophies was a "mirror book," a kind of twist on the pop-up concept. One page had a drawing on it with a piece of Mylar glued into the picture. If you looked into the mirror at the correct angle, a drawing on the opposing page would complete the picture. David tried to sell this book but was told again and again that it was too expensive to make.

The other invention of note he called Vanity Laces. The idea here: a special flap that crossed the front of a running shoe perpendicular to the row of eyelets. This flap had extra eyelets in it running eight across and three high. If you had really long shoelaces, you could thread them up the regular eyelets and then, when you got to this flap, you could run the laces under and over the flap in different directions, creating patterns or even words.

David imagined the Vanity Laces would work for a children's shoe, maybe as a weird approach to literacy, and possibly a fashion item for gangs. This invention he actually tried to patent. He got a backer, a marketer, a patent lawyer and so forth, spending close to $6000 before discovering that his lawyer had done an all-too-hasty search of the database in Washington that records—and has recorded since the Patent Office opened in 1790—what is and what is not new under the sun. Much was missed in his efforts. The claims on the application were too broad, the lawyer told him; at least six other patents had already covered the territory.

Vanity Laces were thus untied forever. But the notion of inventions, and of the elusive patents that allow them to exist as *inventions* rather than rediscoveries of the wheel, had gripped David firmly. From the start of the column, he had an almost completely free hand. (After a few years, we were able to afford him a fact-checker—not that David, with his copious notes and musician's sense of order, ever had much cause to worry for one.) It was Russ's notion, supported first by Phyllis, and for the past three and a half years by me, that "The Patent Files" find its inspiration, its subject matter and style, in David's plain desire to tell stories. We would not direct him otherwise. This was all to the good. The paper's freewheeling, almost libertarian editorial policy matches in some great measure the tenets by which all inventors must work: unfettered by the strictures of convention, outside the lines. Felicity, then, once more.

David almost immediately gravitated toward profiles of inventors, and met with them face-to-face as much as possible—in coffee shops and city parks, at fast-food franchise restaurants, sidewalk cafés. It is the virtue of the profile, he said, to bring out a story that would otherwise be unknown, as opposed to riffing opinions off known news stories.

The interviews, as you will see herein, are for the most part joyous affairs. Unlike most kinds of professions, the inventor's relationship to the media has not already been engineered by handlers, those unctuous P.R. representatives and publicity agents who plague our profession. Contrast a man at home with an invention—a reverse watch, for instance, measuring not what time it is, but how much you have *left*—with a rock musician and his recently released CD, or an athlete with his stock do-the-best-I-can sound bites. In the latter cases both sides know exactly what the questions and answers will be. But inventors do not belong anywhere and do not have many people to talk to. Consequently, they never shut up. Plus, there is no predicting where their minds are going to go.

In this respect, they are very much like David himself. The column evolved quickly, however, and before long it was clear that David had learned something about where *his* mind was going to go: almost from the start, "The Patent Files" illustrated a particular worldview from the vantage point of invention. A particular and wondrous worldview at that: it was David's feeling, writ large across nearly every column, that inventions and the process by which they come to be provide a small and powerful lens to train on . . . damn near everything in the universe.

Within this book, culled from the more than 200 pieces David contributed to *New York Press*, starting with the first column and ending with the last, you'll see genial polemic and historical takes and various other musings on the events of the day and of the days past, all cast in the bright light of *Eureka!* discovery.

David has in the past described himself as a "curious witness," and the phrase fits him well. But it was a friend of his who pointed out that David is more than that by far: the poet of the inventing classes. I think this is really the best compliment, and the truest, that one can pay him.

Sam Sifton
Managing Editor, *New York Press*
Fall 1998

Acknowledgments

MANY PEOPLE DEVOTED their energies to the making of *The Patent Files.* Russ Smith was the rare newspaper editor who not only saw the value of a column about inventors but acted accordingly. Sam Sifton took the lead from there and did the outrageous thing of letting me say what I wanted to say, stepping in only when I needed to be saved from myself. Farther down the production line, Lisa Kearns and Adam Mazmanian added luster to my sometimes tarnished prose and—no small feat—made the process a pleasant one as well.

Steve Roen, patent lawyer and curious person at large, proved to be an invaluable resource as the column progressed, as did countless other friends on numerous occasions. My deepest gratitude, however, must go to the inventors themselves for being so generous with their ambitions, their wit and their time. When Lilly Golden and Chris Pavone imagined *The Patent Files* as a book (an act that deserves an acknowledgment all its own), they were really recognizing these people's work coming up through mine. And so should you.

Prelude

NOTHING IN MY SHELL of a world stays put. Tragedies turn into trial runs. Jokes transmute into lifelong obsessions. Like Oscar Wilde stumbling drunk through an echo chamber, I look on helplessly as my life imitates art and back again.

Which, I suppose, is how I found myself in hot pursuit of a patent. My invention first presented itself to me while I was writing a futuristic short story that fizzled. Then one day, long after I had shelved the first draft, my muse slapped me around a bit and told me I could actually make the thing. *Thwack!* Oscar brains himself on the chamber wall.

Thus inspired, I went forth with God. More than a few people would say that I went forth with foolhardiness. Still, I managed to enlist the forces of a marketing rep, a patent attorney and a financial backer (to the tune of $6000) before the whole caboodle collapsed.

I won't go into the details of my resounding failure. This task I leave to the stenographers poised at the gates of apocrypha, where all the best stories reside. The point is that, in striving to own a small piece of my native intelligence, I became fascinated with the world I had entered.

No, that is putting it too mildly. I wasn't fascinated. I was *hooked*.

As the corporate world would have it, patents entitle their holders to sue the competition. That's true enough. But to my mind, they also make up a kind of encyclopedia of the collective consciousness. The patent files embrace quirks, poems, late-night jokes, brilliant flourishes and shimmering nightmares. Trash bags for use in outer space, coffin alarms for the prematurely buried, the simple light bulb, recombinant DNA—it's all on the record.

Then, too, the very notion of a patent is an invention in its own right. No such beast existed—not in this country, anyway—before 1790, when Thomas Jefferson brought the system into being as a way of encouraging the ingenuity of the common citizen. Two hundred and some years later, the American patent system is the intellectual property of about 180 examiners who sit in a building down in Washington and decide once a week what is and what is not new under the sun. That in itself seems odd and worth considering.

Each week, I'll be using this column to discuss patents on the cusp of their issue and various related subjects. In this overinformed age, I certainly can't hope to encompass the sum of all innovations, nor would I recommend such an approach, but there remains the prospect of building a hardy web here and there.

AS A NATION, we have this overriding obsession with how we smell. Roll it on, spray it on, wash it off—an American can be the most despicable character this side of Dante's inferno and still receive the ovations of his peers, but exude just a whiff of the old b.o. and he might as well consider himself a Third World citizen.

Well, all that could change if Erox Corporation gets its new product up and running, because the studies backing up its R&D suggest that it's not what you smell that counts. It's what you *don't* smell.

Nocturnal television viewers, the kind with a special place in their hearts for the tinsel tease of paid programming, may have already heard of Realm, Erox's fledgling product. A half-hour infomercial extolling the virtues of the new scent has made its first appearances on WOR-TV in the past few weeks. Early birds may have noticed another spin on Realm in the *New York Times,* which ran a story fleshing out the scientific background (September 7, 1993).

Now typically, paid programming doesn't do much for me besides make me wish for a late night movie starring Charlton Heston. But Realm contains pheromones, the chemicals that would seem to govern sexual attraction, and the man who has isolated them (from the scrapings of leg casts, no less), a certain Dr. David Berliner, has filed for a patent on their use. If Realm is for real, it's a secret weapon beyond compare. After all, harness desire and you can rule the world.

The infomercial is peppered with the testimony of ordinary Joes and Janes, describing how Realm has boosted their romantic life. The CEO of Erox, Pierre de Champfleury, makes an appearance, looking very French, and there are shots of horses and damsels in meadows galore. The sum effect, in fact, is to stun you into submission as the scientific data go by, so I'll try to summarize the technical matters.

Pheromones have long been known to guide lower mammals in the selection of sexual partners, among other things, but they were generally thought to disappear in humans before birth. Now, recent experiments have pointed to a connection between these substances and the human nose. When someone exudes pheromones within range, you feel, rather than smell . . . something. Or so goes the theory. Because until the patent is issued, nobody can corroborate the experiments that link pheromones with human emotions.

It also remains to be seen whether Realm uses synthetic pheromones, which *imitate* those found in nature, or whether it uses actual genetic material derived from a human source. For now, the makeup of Real remains under wraps. But if the latter is the case, Erox is bound to be caught up in the biggest controversy raging in the patent community today: whether human genes can be owned.

Fair enough. I knew the jury was still out on this product, but I also recalled that Thomas Edison, the least insane of all the great inventors, exposed himself to about a zillion curies of X-ray radiation in the experimental days and still lived to the age of 70. So I decided to try Realm firsthand—even if it meant applying cells replicated from some ski bum's leg to my neck.

After two days, I would say the results were neither scientific nor overly romantic. A waitress at a diner I frequent was actually perky in taking my order—not a common occurrence. Two women at a brunch gave me a bit of their undivided attention. And in Washington Square Park, I felt the eyes of a few *men* fixed on me. All in all, I couldn't say for sure what any of this proved, except that the fragrance accompanying these Stealth chemicals smelled pretty good.

Not that I had hoped to achieve precise results in the first place. My own prior knowledge of Realm's putative powers may have given me the confidence that produced these vague results. And that might be good enough for a lot of people. The placebo effect commands some respect even within the medical community. But let's assume for the moment that Realm does work, if not in this incarnation, then in the next. Once it's been tweaked to maximum effectiveness, it will no doubt be the answer to many people's dreams. Dealing as it does with the spoils of seduction, it's also bound to raise some forehead-wrinkling questions.

Say a prosecutor wears Realm in the courtroom while presenting his case to a jury. Would such an action predispose the jury to nail the defendant to the wall? Could the prosecutor defend his use of pheromone perfume as freedom of expression? How about an undercover cop posing as a prostitute? If she wore Realm to lure in a john, would it constitute entrapment?

Then there is the increasingly nebulous territory of date rape. How many frat boys will find justification that a woman was wearing Realm and thus leading them on? And how many women on college campuses will claim that any man wearing Realm is trying to rape her?

Dr. Berliner is of the opinion that each person emits a unique mix of pheromones. What, then, are we to make of a product containing a single mix of pheromones for all women and another for all men? Does Realm herald an age of standardized mating patterns?

Questions, questions, questions. Of course—and people will think I'm being ironic here, but I'm not—if the right people got their hands on the licensing rights, the principle of seduction could be taken to a whole new level. Why not introduce pheromones into air conditioners and central heating systems? Walk-in traffic would be enhanced by untold leaps and bounds if pheromones were blown out the exhaust systems of retail stores.

While we're at it, why not seed the clouds? Entire populations would experience an entirely new sense of confidence and well-being every time it rained. Realm could soon become so ubiquitous as to make all discord moot. People would lose their jobs and still feel all right about it. The economy would go down the drain without so much as a blink from an attractive populace.

And at that point, in the relentless logic of our science fictional existence, we will have achieved our messianic goal: our quality of life will have descended to the status of a Third Word nation, but goddammit—*we won't stink.*

September 29 – October 9, 1993

1

AROUND AND ABOUT THE PATENT GAME

A Word to the Wise

GUILTY, AS CHARGED. Over the past few months, I have paraded a slew of inventions before you and yet, for reasons I cannot explain, I have ignored the ceaseless appeals to my considerable (and somewhat instantly bestowed) expertise. But no longer. Spring is here and it's time to throw these letters away. Fortunately, I can respond to the great majority in a single sweep, since nine times out of ten they center on the tremulous question, "What should I do if I have an idea?"

Well, the answer is simple: sit for a moment and hope your condition passes. Many theorists (whose names are ridiculous, I assure you) assert that the world has descended into its current squalor precisely because people insist on promulgating their thoroughly useless notions. Think about it. Aren't there enough ideas languishing in obscurity already? Couldn't you just coast along on somebody else's idea, say, unconditional love, for a while?

Indeed, there are many reasons to forgo the perils of the patenting process. The on-liners, as is their effusive wont, are even predicting the death of intellectual property law in the face of the oncoming information bonanza. How, they ask, can you ever hope to protect your idea in an age when anyone can upload your records and make them available to half the world in the blink of a gigabyte? John Perry Barlow entered this subject with guns blazing in the March 1994 issue of *Wired*: "While there is a certain grim fun to be had in it, dancing on the grave of copyright and patent law will solve little, especially when so few are willing to admit that the occupant is deceased, and so many are trying to uphold by force what can no longer be upheld by popular consent."

Never mind that the enforcers of whom Barlow speaks are carping mostly about software rights while the patents for dental floss, propellers, oil rigs and the rest of the carbon-based world roll merrily along. Barlow's argument is still spot-on: patent law is growing ever more nebulous, and the lawyers are growing fat rings in the process.

All right. If someone else's fat rings don't bother you, the first step is to write your idea down. This stage is kind of fun, but you will see that already things are getting more difficult. How, you must ask, is your invention different from all other inventions? To find out, you must do your own preliminary patent search. A more extensive search will have to be conducted later if you decide to proceed, but right now, you are trying to rule your invention out. Go to your nearest Patent Depository (in New York, the depository is in the process of relocating to the Manhattan Midtown Library on Fifth Avenue and Fortieth Street) and do everything possible to *disprove* the originality of your idea. If you're lucky, you'll find what is called prior art, meaning someone else has thought of it first.

But say you really do have something new. How can you be sure it will work? Many inventors find the question annoying, so they never bother to make a prototype—until it's too late. Take myself, for example. I made my prototype only after the marketers, the lawyers and a friend who offered to back the idea were completely onboard. You can't believe how discouraging it is to struggle with an ugly, non-functional model when there's skin in the game.

And so a nerve-wracking situation develops. You must keep the great miracle utterly to yourself, yet it looks like you'll need some help. Technical prowess may be required to build your prototype, for example. Or maybe you need to find someone who will pay for the securing of your patent.

One school of thought here is to do as much of it as you can yourself. You are not an idiot. Or at worst, you are only selectively idiotic.

For examples of self-sufficiency, take a look at the people reviewed in these pages. Geoffrey McCabe saved his money by writing all of his patents himself. John Walter is marketing his True Mirror all on his lonesome. Iman Abdallah actually became a patent attorney to further his own inventions. Sure, the world is big and scary, but after all, it's your idea, so you have a say as to how it gets out there.

The other school says you should generate ensemble excitement. If you can get a marketer or a backer interested, the theory goes, there must be at least a smidgin of demand for your invention. Those who go this route, and I did, enter the giddy world of disclosure. Make sure absolutely everyone, even your mother, signs a non-disclosure document, which is a written agreement that the other party won't tell anyone else about your idea. They will, of course, but at least you'll have a trail of accountability. If you're exceptionally paranoid, you might even try giving away a bogus idea, just to see if the lid stays on.

Lawyers are not hard to find—far from it. Freelance attorneys tend to be preferable, though, because in the parlance of law, you are a small entity, and in the parlance of truth, so are they. My lawyer, a big-firm type, consigned my search to a low-level drone who missed some examples of prior art. This little mistake effectively grounded the entire operation some $6000 later, when the Patent Office sent back its rejection.

Rough waters, for sure. Still, the game thus far is a snap compared to the selling phase. Marketers—the people who will try to sell your idea to a manufacturer—can be found through word-of-mouth, as in, "Do you know anybody who markets weird ideas about filing cabinets?" They also tend to show up at trade shows related to their field, so you could nose around a few of those. No matter how hard you try, it will be hard to tell the crooked from the straight, because the marketer's rap is smooth as replicated skin. Any intelligent marketer will want a percentage of the take, but you should steer clear if they want money up front. And avoid the kind who advertise on the radio or in the classifieds. They're trolling for suckers.

In trying to charm your way into the marketer's hummingbird heart, it is worthwhile to remember that everyone is full of shit, yourself included. Play lots of poker in preparation for your meetings with these quixotic, unreadable people. Learn to bluff, and to read the slight quiver of lips. Learn to understand that when they have left the premises, they aren't playing anymore. Maybe.

By now, unconditional love is probably looking pretty good, but suppose you manage to make your way through this mine field. You patent your invention, and somebody agrees to manufacture it. Maybe you even get a good royalty deal. And *maybe*, in defiance of statistics, the thing actually starts to sell. So you're in the clear, no?

No. A patent does not give you the exclusive rights to your invention. A patent merely gives you the right to sue those who manufacture your invention without your consent. So what happens if a few big lugs decide to disregard your inventorship with extreme prejudice? Don't you think they can clobber you with legal fees until you relent and go hobbling into your honest but not very well-lighted corner? Yessir, that's exactly what they can do.

Not exactly an encouraging scenario, but at least you know what to do the next time you're in one of those arguments about who thought of something first. Smile your best lizard smile, let your eyelids go lax—and defer. Believe me, it will add years to your life.

April 6 – 12, 1994

At the Patent Office

EVERY NOW AND THEN, you'll see an article published about the U.S. Patent Office. Typically, the article will mention the founding in 1790, the 180-some examiners, the 10,000 patents that issue forth from its halls every year. What they never tell you is that the Patent Office is housed in a glassine rocket that never takes off.

The doors slide open, letting us out onto the jarringly clean platform at the Crystal City stop, in Arlington, Virginia. Paralegals and linebacker women scatter to their appointed rounds. We're underneath the turrets and domes of a self-contained community, in the catacombs of suburbia.

Jarring is a subjective term, of course. People who visit New York invariably describe the impact, but for myself, I always feel the biggest shock when I leave. Beyond the city limits, the outlines of objects are crisper, the colors brighter by many orders of magnitude. Pedestrians don't blend nervously into their scabby surroundings; they stand upright and maintain their identities. Surely, they will notice me. Surely, they will ask for my papers.

But they don't. We walk unnoticed down an endless underground corridor, past cut-rate clothing stores and spanking-clean supermarts, past bookstores and au bon panderers, decompressing from a ride on the safest subway in the land, which cuts through the most murderous city in the land (and maybe all of white D.C. is a spaceship at that) until we arrive at a placard that reads, "Public Search Room—CP3-1A03."

Hmmm. CP3-1A03. Yes, I know where that is. Still, I'd like to see if the delivery man coming down the hall knows, too. He frowns, then shrugs and tips his cart back, the weight of it catching against his hands. No sooner is he gone than an immensely freighted woman, entering my private quiz show with a set of mammoth lungs, hollers from the open elevator door: "Search Room, honey? Come with me."

And they said *21* was rigged.

The Patent Office may be squirreled away among various spandex fiefdoms, but it does manage a serviceable flourish of the old-time pomp. Outside its doors is a series of display cases, each paying tribute to a patentholder who has made this country et cetera. Hidden among the standards—Pasteur, Marconi, Shockley, Eastman—there are even a couple of surprise cameos. Robert Goddard, pioneer of rocketry (and godfather of Crystal City perhaps) gets a nod, even though he was reviled in his time, as does the underdogged genius, Nikola Tesla. Naturally, though, no showcase is as gussied up as the one reserved for that most mendacious of inventor-icons, Thomas Edison. In fact, the Wizard gets the only attempt at state-of-the-art, with a button that implores, **PUSH TO START.** I pushed. Nothing happened.

To the side, another case contains models that were submitted for the examiners' approval as long ago as the twilight's last gleaming. I looked at these faded artifacts for some time without being able to divine their function. Were they reapers? Riveners? Shriveners? Never mind the strong argument to be made here that the past is more of an Other than any demographic on the planet today. It's a spooky display, with just the right mix of dankness and Yankee doodle dandy.

No such curios are evident once you're inside. In the Patent Office, as in any of the patent annexes sprinkled around the country, the business of acquiring intellectual property is a quiet one. Stacks of documents gather the smells of nervous men. Underemployed hopefuls hog the CD-ROMs. Lawyers pore over precedents, trying to squeegee an original thought from their client's boasts. What *is* different is the access to the files. Only here will you find the complete record of every

U.S. patent ever issued. And only here can you learn for certain whether anyone else has snared your idea before you, in a process known as a search.

The importance of the search is a subject I know a thing or two about. Two years ago, I tried to patent an idea that, if odd, at least seemed novel: a method of spelling out words by stringing shoelaces through a grid of eyelets. The idea was compelling enough to lead me to a lawyer, who performed a search for about $900. Soon thereafter, I was informed that there were no patents resembling my idea, so we went ahead and filed an application. This part of the process ultimately cost me and my backer—a friend of mine with the lucidity to throw money out of my particular porthole—an additional $4600.

Sad to say, the patent examiners rejected my application after unearthing six patents that were awfully similar to the one I proposed. My lawyer's gofers had simply fallen down on the job. Too bad, tough luck, goodbye $5500.

Well, now here I was, with two eyes, four limbs and a brainful of passable neurons. How long, I wondered, would it take me to find those six patents, the ones my friend has paid so dearly for, given my amateur status and my ignorance of the turf?

I looked at the wall clock. *Ten minutes to four.*

The first hurdle was the directory. This tome is something like *Roget's Thesaurus*: you proceed through a maze of classes and subclasses toward the theoretical area your invention might someday call home. There are more categories than you would care to imagine. When I sat down at the directory, for example, it was opened to "Spittoons; trapped liquid, gravity receptacle."

I moved with reasonable haste. After picking out "Class 36—Shoes," I ran my finger down the roster of subclasses, pausing with due respect at "Subclass 19.5—Cement only," before landing on the best bet: "Subclass 54—Tongue pieces." From there, I got waylaid by a computer, which shimmered with inscrutable numbers while a Patent

Office employee joked with a regular and ignored my pleas for help. In fact, I had given up and was leaving the premises when I noticed a list, lying unmarked on an out-of-the-way table, that told me what I needed to know. Bingo. Up one flight of stairs and I was there.

My quarry lay in a cubbyhole, exactly as marked. No one seemed to be monitoring the stacks at all. With the reverence of a choir boy in the rectory, I pulled out a ream and sat down beside a few clerks who were telling inside jokes and disdaining the passage of time. I scanned the first few patents, and there it was: my idea with someone else's name on it.

I kept going. After combing only half of the pile, I had found 14 patents that, two years ago, would have sorely discouraged me from forging ahead. *Fourteen.* The patent examiner, who identified only six, must have taken a few potshots, realized my application was nowhere and gone to lunch.

With pursed lips, I straightened the pile and returned it to its cubbyhole. Then I strolled downstairs. *Eleven minutes to five.*

In less than one hour, some low-level drone had spent more than five grand of my friend's money. If that's not a cautionary tale about the evils of lawyers, I don't know what is. Still, instead of making me depressed, my belated coup had me feeling almost giddy. I had labored against the machine and, though the cost proved terrible, I had beaten it. Never again would I feel put off by the men in gray.

On my way out, I stopped in front of the Edison display for a final look. Edison, stealer of patents, beacon of thieves. I pressed the button to start again, and again nothing happened. There was something inexplicably satisfying about that.

November 2 – 8, 1994

The Inventor as Terrorist

FOR SOMEONE LIKE ME, who spends so much time ferreting inventors from the woodwork, it's an odd experience to watch Theodore Kaczynski come into the light, dry eyes blinking. What only yesterday was a nondescript cabin near Lincoln, Montana, has suddenly become a diorama of mechanical effects, each one more fascinating than the last. The headlines block out Kaczynski's iconic moniker in 32-point type, and the articles under them chart the clues to his trade to the tiniest detail. Even the intricacies of typewriter construction are dredged up from the morgue.

I guess you could say I'm jealous. The FBI and attending press have been able to rout out the most glamorous inventor—sorry, make that suspected inventor—of our time, while I'm left at home, waiting for the next bean-bag genius to call. So thorough has their research been that I'm tempted to think that the feds are planning to reverse-engineer Kaczynski's gadgets and go into production for themselves.

A weird comparison? Arch? Far-fetched? Maybe, but it's worth asking where the terrorist leaves off and the inventor begins, especially when you consider that history has lauded many a man who would beg the question today.

Granted, the question has been a long time coming. Many inventors of the early republic exhibit a fondness for making things explode. Oliver Evans built a gun before he reached puberty; Count Rumford worked up fireworks as a teenager. But all this was dismissed as tomfoolery in its time. Only on April 5, 1996, did the *New York Times* manage to work a similar detail into the profile of a terrorist, with the remark that, as a child, Kaczynski "built small bombs and blew up garbage cans."

The inventor-terrorist was still nowhere to be found in the 1850s, when Cyrus Field was trying his damnedest to lay a telegraph cable across the Atlantic. While going about his business one day, Field was stopped by a local official who asked what he had in his bag. Field produced a length of thick "wire rope" and declared, "This is a piece of the Atlantic cable!" This length of wire rope was not identified as material that "can be used to provide power for a device used to detonate explosive material." The interlocutor, who had expected to find a wad of counterfeit bills, apologized for his error and let Field pass.

In fact, it took newspaper stories with real eye-popping appeal to get the idea of the inventor-terrorist going. Consider this passage: "Most ghastly was the sight of the mutilated corpses strewn on the ground. Not only had the clothes been torn off but on some the head was missing and the flesh ripped off the bones. These formless masses of flesh and bone bore little or no resemblance to the human body... [One] unfortunate victim was still alive and carried to the hospital on a litter, looking more like a bloody mass of meat than a human being."

This was not a report from the trenches of World War I. It was a description of the scene in Heleneborg, Sweden, on the day when Alfred Nobel's entire dynamite factory went off by mistake. That event, horrible and unforgettable, dogged Nobel for the rest of his life. Dubbed the Merchant of Death, he responded by paying the world back in prizes.

Of course, there's a difference between dynamite that goes off inadvertently and a Molotov cocktail hurled into a phalanx of coppers. One is a regrettable error, the other an act of intent. But as history progressed and inventions grew increasingly dangerous, the question of intent only got thornier. In the 1890s, General Electric was established on the strength of its alternating-current technology, at the time considered potentially lethal. GE then set up the first R&D lab in the U.S., setting the pattern for what became the American century. It is no coincidence that this happened just as technology was running way from its creators; corporations, after all, are designed to limit liability.

Nor is it any accident that, as the GEs and GMs of the world flourished, the iconography of the "lone" inventor started to kick in. Early on, the press covered figures such as, say, the young Guglielmo Marconi, as reclusive eccentrics, whose lairs revealed a dizzying array of wires, bottles and manuals. Later, when the press no longer covered them as celebrities, disposed inventors started writing long credos based on their own personal cosmologies. Buckminster Fuller first went into action as a serious inventor with *4-D*, a "meticulously handwritten" manifesto that made short work of bankers and industrialists. Similarly, Wilhelm Reich raged against the system in his tract of priceless title, *Listen, Little Man!* Loners with gadgets in their garrets, writing prose against the dying of the light . . .

And so here we are. No doubt the Unabomber manifesto can be read as one more screed against the industrial complex. Burn it down and all that. Yet a guy who spends his days keeping his switches in good repair has got to become handy with tools after a while. Clearly his complaints are not with technology as a whole; he's been tinkering away to his heart's delight for 18 years. There are nuances here, twitches in the psychological profile.

A recent *Dateline* segment suggested that Kaczynski was obsessed with his typewriter, and went on to point out the irony that this machine is in fact a crowning achievement of the Industrial Revolution. Personally, I don't think Kaczynski was obsessed with typewriters; he simply had to write his 35,000 words on *something*. But add that trusty bicycle (first made practical in the early

1890s) and a few piles of *Scientific American* to the myriad electrical experiments and he does emerge as a late-19th-century man.

One hundred years ago, Theodore Kaczynski would probably have become a well-known inventor. Given his predilections, he probably would have forged some kind of device that threatened the commonweal—say, the machine-gun, or an x-ray camera. But we no longer live in a world where loners are aggrandized. (Take it from a writer. Solitude is totally out of sync with the way this country works.) So Kaczynski, assuming that he is the Unabomber, became an omega inventor instead, railing against the cumulative effect of his forebears' machines.

You have to admit that it's been a highly successful strategy, since Kaczynski now occupies the same rarefied position in the public consciousness that 19th-century inventors once did. When all is said and done, he's probably the first inventor to fetch a week's worth of top news stories since the Wright brothers' day.

The ironies are almost too thick to bear mentioning. Having set out to build a new world, inventors have been suffocating under its weight for some time now. In a sense, it's only logical that they would try to take it apart again. (As they say, take something apart and put it back together often enough, and you'll end up with two of them.)

Anyone who undertakes such a job is still in for a lot of work; our conglomerate theme park is not likely to suffer dismantling without a fight. In the meantime, I have only one sick consolation: as long as the world prevents inventors from staking their claim except in the most ineffectual corners, I'll have plenty of time to scoop my competitors. There will be more Unabombers to come.

April 17–23, 1996

A Good Christian Lawyer

RECENTLY, I ADMONISHED inventors to forget about their own ideas in favor of the perfectly snazzy ideas already out there, ideas like—oh, say, brotherly love. Since that time, of course, brotherly love has broken out all over, which just goes to show how much power a journalist really does command. And yet if I'm to obey the Golden Rule, I must admit: Skip Throgmorton was onto the trend before I was.

It began, quite naturally, with *Popular Mechanics.* Leafing through an issue in search of the Heath Kits of my apocryphal youth, I spotted an unusual ad. "Christian based inventors Co.," it read, "Patent Drawings, Patent applications as well as Sales. Throgmorton Crest." The area code indicated an Arkansas address.

Well, I thought, seek and ye shall find, so I dialed the number and listened to a deep Arkansan drawl announce itself as Skip Throgmorton. (A friend of mine has taken to mumbling *omen est nomen*—"the name is destiny"—lately, and maybe he's onto something. The coincidence of people with unorthodox names and people with unorthodox lives is simply too overwhelming to dismiss.)

Throgmorton, as I soon found out, began his company after having been thrown to the lions. In the early '80s, he came up with a board game called *Universal Star*, which employed a Monopoly-like board and cards containing information about the different planets. The object was basically to amass more landing pads and spaceships—more *stuff*—than anyone else. When he reached the patenting stage, Throgmorton took the game to a sales company, and the next thing he knew, it was on the market.

Only problem was, no one had bothered to give him credit as the inventor. Incensed, he contacted the company that put the game out, only to be given the brusque boot. *Their* inventor, they said (with some canniness), had seen a vision in the wilderness. That about put an end to the argument—communion with God is a tough defense. Several inventions and several tough defenses later, Throgmorton grew weary of the betrayals and decided to take matters into his own hands.

Throgmorton Crest is not exactly a super-slick organization. It consists of one man sitting at a desk in Pocahonts, Arkansas (pop. 6000), tapping away at a 486 computer. Throgmorton doesn't even have the cachet of being a patent attorney. What he does have are high ethical standards. He will steer inventors clear of scam operations, like the one that absconded with his board game, for example. He will also make drawings, recommend an attorney if it's called for and generally advise his clients. But his real good deed is the fee.

"I gotta sleep at night," he told me. "If an inventor comes to me, I charge $50 to assess the invention, unless it's already patented. If it's patented, I charge nothing. A regular company will charge $200, $250 for the same work. Well, you can sit down and write a report and tell somebody a bunch of bullshit and earn your $250. I've seen their reports. I know."

Given his fiscal behavior, I had no doubts about his honesty. But I was still in the dark as to what a "Christian invention" was. Did he have limits as to the kind of proposals he would take on?

"Oh yeah," he replied. "If some clown wants to send me a dildo, they can keep it. Basically anything to do with pornography, I'm not interested. I'm not talking about condoms, though. I think they're still up in the air."

Pressing for finer resolution, I asked what he thought of Patent No. 5,080,621, issued to Alan W. Nayes of Orange, California, in 1992. This patent covers the claims for a Water Walking Device—

two buoyant hulls with propulsion flaps on the bottom. Could a Christian-based invention company endorse a device for walking on water?

"You mean to deceive people?" he asked. "Or as a kind of a novelty device?"

I thought of Orange, California. I thought of Hollywood, guava-burgers and bladers in bikinis. I flipped a coin. "I think it's a novelty device," I said.

"Oh, I don't have any problem with that. I think it would be kinda neat. I don't have any problem with people having fun. But I'll tell you what: if I was involved with a ministry and the minister was using them, I'd knock him down. Right in front of everybody."

Not your usual portrait of a dogmatic Christian fanatic. Indeed, Throgmorton's quickness to imagine a minister in sheep's clothing extended to a disaffection with the religious right in general. "With religion I am sick," he proclaimed, and went on to describe the fractious atmosphere around him: Protestants condemning Catholics, the Church of Christ condemning Pentecostals, a whole Babel of evangelists working their perfidy. To hear him tell it, something is rotten in Dixie.

Though internecine skirmishes have caused Throgmorton to despair of churches (he now restricts his religious observances to Bible study), he still believes in the necessity of being born again. He described his own conversion in terms surprisingly void of rhetoric, casting himself as a sort of Job-as-Joe. He didn't sound like a preacher. He sounded like an inventor who happened to have some theological buttressing at his disposal. Who knows, he might have whipped a good fire-and-brimstone sermon if I had kept him on the line long enough, but it was not to be. The interview came to a close with a commotion in his office.

"Excuse me, someone just came in," he said. Then he added brightly, "That's okay. He's a Christian, too."

In the course of our conversation, Throgmorton had insisted that certain passages in the Bible referred to invention, even if the specific chapter and verse eluded him. Piqued by the notion, I scoured my gold-leafed Gideon for gadgets, hoping to conduct a massive technological exegesis. I didn't get very far, but I did learn a thing or two.

Aside from Paul, who basically invented Christianity outright, the New Testament is curiously devoid of accessories. This makes sense, of course. In a world of hucksters and conjurers, the Son of God isn't going to need props except as metaphors for affairs of the heart. Sure, he turns loaves into fishes and water into wine, but only once. It's not like you could start an industry with that.

The Old Testament, on the other hand, tells a detailed story of who invented what. In the beginning, humans had the field to themselves. The very first invention, cited in Genesis, is the apron that Adam and Eve patch together after eating from the tree of knowledge of good and evil. From there on in, the people get gizmo-happy. Expelled from Eden, the first couple bears a tiller, who uses a till, and a shepherd, who, I suppose, uses a shep. The following generations come up with tents, musical instruments, metalwork and cities. Not until God tells Noah to build an ark, in fact, do we see the first invention inspired by supernatural forces.

The instructions here are very specific. "The length of the ark shall be three hundred cubits, the breadth of it fifty cubits, and the height of it thirty cubits. A window shalt thou make to the ark, and in a cubit shalt thou finish it above; and the door of the ark shalt thou set in the side thereof, with lower, second and third stories shalt thou make it."

Not only does that sound suspiciously like a patent application, but I'd be willing to bet that some canny company made up the bit about Noah's communing with God, too. As I said, it's a tough defense.

August 3–9, 1994

Patenting a Gesture

GESTURES ARE IMPORTANT. We know this because the world is full of half-considered gifts. But I had forgotten that gestures could be lucrative until Dr. Samuel Pallin, a physician working in Phoenix, Arizona, came along to remind me.

Dr. Pallin, you see, has patented a method of cutting the cornea. Instead of making a normal cut, which requires stitching, he makes a V-shaped cut. This V-shaped cut apparently needs no stitches; it heals by itself. Good idea, and maybe Pallin would have left it at that, but when somebody else started using this V-shaped cut on his own patients, it sort of, well, bugged him, so he took out a patent and threw a lawsuit at the infringer. Not that he wanted to get rich. It was just a gesture, you might say.

I happen to know all this because I was listening to National Public Radio's "All Things Considered" a few weeks ago when Pallin faced off against George Annas of the Boston University Medical School. Annas did not like this V-shaped-cut patent one bit. In fact, he charged Pallin with deprofessionalizing the medical industry. Doctors come up with new techniques all the time, he grumbled, but they don't go around getting patents on them. Pallin, sounding hurt, sallied back that four out of five doctors are perfectly happy to jam their offices with patented gizmos and then set them loose on their insurance-laden patients. How, then, can the same people turn around and complain about a patented technique? Some of his colleagues, he said, had actually urged him to scrap the technique business and come up with a patentable tool instead, even though he didn't need one.

At this point the moderator cut in, asking about the Heimlich Maneuver. Was that patented? No one seemed to know. I don't know, either, so the next time you save a choking person, you might want to do a quick assessment of your savings account first. I do know that I come across such things in the patent lore fairly often. No lie: not long ago, I noticed a patented massage technique, sitting there in cold print.

Still, we should take pause, because deprofessionalizing the medical industry is a strong claim. After all, there's so little deprofessionalizing left to be done in the medical industry that anyone who claims to have found another way has got to be a liar—or a genius. In fact, as I thought about it, the possibility that Pallin was a genius started to seem sort of exciting, so after the show was over, I went searching for patentable techniques in professions that hadn't been completely deprofessionalized yet.

This was hard work. My own profession was totally shot, of course, except for a few obvious oversights. I could start to completely split infinitives, for example. Unfortunately, no one really benefits from a split infinitive except for people with cut-up corneas, who are happy just to be reading. Besides, a split infinitive isn't really a technique, unless an editor slips one into a carefully written manuscript.

No, the universe is narrower than that. Physical motions, gestures—these are the ticket to patent glory. I should add that a patent does need to have a function. That's what the Patent Office says. As luck would have it, though, the idea of function is wide open. All you need is a reasonably plausible explanation for an insane proposal, which, if you watch the news with half an eye on any given day, should be no problem. I mean, Channel 4 recently ran a news story about scientists in London who studied why toast always lands with the butter side down. Were those scientists running their tests just to get on TV? No sir. You can bet that an I-Can't-Believe-It-Landed-Butter-Side-Up brand of bread, or butter, or pre-buttered bread is in the making as I write.

Anyway, once I got the ground rules down, I thought of several techniques sure to deprofessionalize somebody somewhere along the line, now that I know well enough to patent them.

Try this one on for size. With the Post Office looking more like the streets of Rio every day, they've been forced to skimp on the adhesive on the back of their stamps. In fact, only a very narrow range of tongue pressure, combined with just the precise secretion of saliva, will now suffice to make a stamp stick to the envelope. But do I know the combination? You better believe it. Vast amounts of research, teams of specialists working night and day, detailed analysis of the tongues of mesomorphs, ectomorphs. You know the drill. So if you've been mailing letters and they've been *getting there*, watch out. I'll find you out soon enough.

Here's another one, more in the educational vein. You go to buy a carton of milk at the corner deli. When the guy gives you your change, he doesn't hand you the silver first and then hand you the bills. He puts the silver *on top* of the bills. Obviously, he thinks you're supposed to hold on to the milk carton with one hand, carefully slide the silver into your mouth and then put the bills in your wallet—with one hand—while he looks at you as if you're a bolt in an assembly line and hollers "next!"

Not with my new patent-pending technique. Just take the mongrel assortment of cash and give it a quick flick—just a tiny, V-shaped flick, actually—just enough to make it look as if you've fumbled it. The change naturally falls into the deep recesses of a dubious-looking Korean-candy bin, and the cashier has to dole out the difference. Do it enough times (you're a loyal customer, after all) and he's guaranteed to figure out that there's an easier way. True, you'll owe me the change when you're done, but the world will be a better place for it.

Of course, there's a certain satisfaction in singling out certain groups with your technique patents. Take the Nazi salute. Why has no one thought of this before? We all know the function of the Nazi salute. It's supposed to professionalize hate. That's why I've got an application pending to claim intellectual property rights for this gesture. I don't mind hate, after all. I just think it should be unprofessional.

As you can see, once you get going, there's simply no end to the techniques that can be patented: handshakes, high-fives, air guitars, air kisses, every move a rapper makes.

One word to the wise, though. If you want to patent certain moves used in knife fights—and I say this whether you're a postal worker, a deli cashier, a Nazi or all of the above—you should learn a little bit about the patents that already exist in this area. Cutting the cornea in a V-shape, as you may recall, is already taken.

July 5-11, 1995

Behind the Infomercial

RULE NUMBER ONE: the pitchman never dies. Revolutions may come, political parties may go, but when the smoke clears, there will always be a dubious character lurking among the remains, ready to sell you the shrapnel. Orient yourself at the scene of the conflagration and you'll hear his song—insistent, grating, unavoidable: *People, I'm proud to offer you this lefthanded windshifter right now, at a special, low low price. Why should you buy it, even though you already own ten? I'll tell you why. This windshifter is unlike any other—and for less. And if you're not satisfied with it, you can keep it as a complimentary gift. That's right! I'm that confident about this miracle of modern science—even if you never use it at all!* . . .

Twisted metal creaks in the breeze. The stench of disaster rises. But you can't stop listening.

Such is the bizarre image that works its way through my mind in the late late hours, as one infomercial follows another on my anemic TV screen. The calamities rage all around us, but apparently the pitch must go on. The abdoflexamite machines must be sold. The miracle perfumes must be proffered. The vacuum-powered haircutter must wow the nation's night owls with its sheer improbability. After all, we can't let tragedy stop us. Goddammmit, we have a *culture* to uphold!

I'm only being partly facetious here. You may think of infomercials as a recent blight upon these United States, but in fact, they've been around forever in one form or another, and for a while, they even managed to be fairly potent events. In another age, that con man in the ruins was apt to come bearing an epochal invention.

You might even say that the precursors of the infomercial are what made this country great.

It didn't seem fated to turn out that way. In the 18th century, American pitchmen were just as marginalized as their late-night counterparts are today. Many of the famed Yankee peddlers made their fortunes selling nothing more Earth-shattering than poorly made clocks. One notable mountebank of the colonial era, Elisha Perkins, gulled George Washington into buying his Metallic Tractors, a tong-like device alleged to draw disease from the body.

So far, the usual nonsense. But while the peddlers and the Perkinses were generally not the kind you'd bring home to mother, their promotional style was eventually adopted by entrepreneurs of every stripe. So successful were their techniques, in fact, that by the 19th century, it was pretty hard to tell the con man from the serious inventor. The itinerant who drew a crowd in Poughkeepsie might be selling hocus-pocus, or he might be selling the first workable sewing machine. Granted, the system was not very efficient. For every hundred shysters, there was one Isaac Merrit Singer on hand to push through a major invention. Still, that's not a bad yield when you compare it to the number of important products sold through infomercials today.

So why did the live pitch produce the odd genius here and there, while infomercials seem capable only of juicemen? The reason comes down to access. In the old days, a plank and two barrels sufficed for a stage, and attracting a crowd was no problem. There was a plenitude of public venues.

Then came TV and everything changed. Suddenly, access was limited and expensive. During prime time, only well-endowed corporations could afford to shell out for commercials, and the pitchmen were forced into the margins again. There, they've been able to keep the mountebank's tradition alive during the wee hours—but only just barely. With airtime at a premium, the opportunities for exposure are few, and the chance that a valuable invention might slip through almost nonexistent.

This is where the efforts of people like Steve Dworman come in. Certainly, if anyone has the background to elevate the infomercial to a higher plane, he's the guy. When he was still a teenager, he invented Lights Out—sheets of black plastic that adhered to windows by static electricity to create an instant darkroom. Soon afterward, he attended film school at UCLA and wrote TV scripts for shows like *Happy Days.* These two interests—media and ingenuity—converged in 1987, when he started what he claims was the first home-video dating service.

But Dworman never cottoned to the video business; the distribution was too sloppy for his liking. Noticing that infomercials were coming into their own, he "put a couple of deals together"—Hollywood parlance for nothing in particular. Then he realized how little information was available on the infomercial industry and started publishing the *Infomercial Marketing Report*, a trade newsletter that sells for an astronomical price.

As it turned out, Dworman got into the infomercial business when the getting was good. Ronald Reagan had deregulated the amount of advertising that could appear on TV and cable was just beginning to take off. These were the glory days, when madcap promoters could buy half an hour of cable time for $100.

Not surprisingly, TV pitchmen were legion back then, but it wasn't to last. By the late '80s, two figures emerged from the pack. Tony Robbins, a motivational speaker, began selling a self-help package called *Personal Power*, using prime-time production values. Around the same time, Victoria Jackson put Ali MacGraw and Lisa Hartman in front of a camera to sell a line of cosmetics. When Victoria Jackson became the fastest growing company in the U.S. solely on the strength of that infomercial, the Fortune 500 sat up and took notice. The cost of airtime went up. The glory days began to fade. Eventually, the unthinkable happened: when a marketing director at MCA Home Video tried to contact a home shopping channel, no one returned her calls.

That gave Dworman pause. If a Hollywood hotshot couldn't get a callback, small-time entrepreneurs didn't stand a chance. Ever quick to spot a hole in the market, he organized a convention, called the Sell Your Product on Television Treasure Hunt, designed to put inventors (or companies) face-to-face with media buyers.

This year, the fifth Treasure Hunt will take place from September 16 to 19, in the Park Hyatt hotel in Century City, California. A maximum of 50 entrepreneurs will be in attendance, at $1695 a pop. The festivities will begin each morning with general sessions, during which time showgoers will be coached in—what else—how to pitch their inventions. Then it's on to the private, half-hour sessions with the buyers themselves.

Is Dworman using his Treasure Hunt to sell nutmegs to pitchmen? If so, the set of criteria each product must satisfy make for a fairly elite corps of con artists. The average retail markup has to be around 500 percent. You need to have a patent (or one pending) and a working prototype. The product should sell for at least $39.95 and have mass-market appeal, with data to back the claim up. And "Stop the Insanity" campaigns notwithstanding, it can't be prevention-oriented.

"Prevention doesn't sell on TV," Dworman states firmly. "People have to be motivated to pick up the phone."

It seems to me that some of the fabled mad inventors knocking around the American landscape could meet those requirements without much fuss. At any rate, I'd be happy to see them try. I'd like to rub my bleary eyes one night and adjust them to the sight of a crazed prospector selling a hydroelectric car, or failing that, maybe a flying saucer. If nothing else, it would take my mind off the cataclysms for a minute.

September 11 – 17, 1996

Inventing On-Line

WRITING IS MORE than saying but less than doing. This makes writing a strange activity, and it makes us writers even stranger. Adrift in the alphabet, we live in a state of fevered commitment to something that isn't quite real. If you ask me how I'm doing, for example, I'm likely to gesticulate and mutter half-truths, because I don't really know how I'm *doing*. I have a fair idea that I'm bigger than a breadbox. After that, I'm lost.

This shadow world takes on a new dimension of weirdness when the writing is done on-line. Take a look at these missives. Conversations take place. Truths are asserted and sacked. Deals are broached, then escorted into private e-mail compartments. What exactly are these conversations? The reporter in me tries to apply the rules of "the story" and falls short. If they're events, they're events with no smell, no color, no tempo. They lack what T. S. Eliot called the objective correlative—the concrete detail—which has been an article of faith for reporters ever since the New Journalists came down the pike.

Inventions, on the other hand, fall much closer to the "doing" side of the scale. Once something's been invented, you can never go back. It's real. It's done. It's an action. Indeed, an invention may be the only proof that things really do change.

With these two extremes in mind, it will be interesting to see what happens to BUG.TXT, a CompuServe thread (in the Library Section of the Ideas and Inventions Forum) that documents an attempt to invent on-line, in full view of the world.

It didn't begin this way. It began when David Giacomini posed an idle question about Chinese cockroach chalk. This piqued my interest, because I'd used this stuff in my kitchen about a year earlier and had rid my environs of roaches immediately, forever. (Chinatown, on the street. Two for a dollar, four for $2.17. Don't ask me.) Apparently, I was not alone. Almost immediately, a slew of inventors jumped in with their opinions. The prevailing view seemed to be that the chalk is made of boron, relatively safe for humans, but apocalyptic for roaches. Seems boron is made up of microscopic blades, which lacerate the innards of roaches and cause them to "bleed" to death.

Well, the conversation went on in this jolly fashion for a while, but then something happened. Eddie Paul remembered a setup he had seen in Mexico once—a ceiling covered with plastic bags, filled with nothing but water. The local wisdom had it that these bags kept mosquitoes away.

An inventor's field day, this one. The responses started coming faster than you can say "baud rate." How? Why? Is the water cold? Are the bags big? Are the mosquitoes Mexican? Soon, a cast of principals developed: Eddie Paul, from California; Allen Dittmer, from North Carolina; Randy J. Fleet, from Florida, and J. V. Cloud, location unstated. One reported that he had been to the library, studied the shape of a mosquito's eyes. Another clocked in with the latest bug-bag experiments in his shop. It was established that mosquitoes seek heat. Variations in ozone levels were explored.

Think about this for a second. These people have never met each other. They have no idea what the other eats for breakfast. They have none of the most basic objective correlatives required for any job interview. Yet they're working together like old comrades—conferring, doubting, trailing off into elaborate jokes. For God's sake: inventors joking with one another. I've been on this beat long enough to know how rare that is. In fact, the public hasn't been privy to this sort of jocularity—wisecracks that threaten to become indistinguishable from physical inventions—since Edison's early days.

Eventually, a cameo player pointed out the obvious—that this was a case of inventing in public—and asked what the legal ramifications were. A lawyer dutifully came on to say that inventors have a one-year grace period, after making a public announcement of their invention, before they have to file. A sysop also chimed in: an edited version of the thread would be saved in the library, she said, but couldn't be used for legal purposes. Each party would have to download as they saw fit. (The thread as it stands is fairly well shorn down. The lawyer's comment, for example, didn't make the cut, nor did a lot of the jokes.)

Meanwhile, the conversation reached ever higher crests: mosquito mating habits, trips to Mexico, death rates of captured bugs. Eventually, I couldn't resist, I entered the fray—as usual, as the dissenting voice. How, I wondered, does the international nature of an on-line conversation affect patenting rights? For this I was summarily rebuffed by Paul. Who knows, he said in effect, until we try?

Fair enough, but I think it's worth investigating where an attempt to invent in public begins and ends. For example, there's the sticky matter of determining who contributed what. R&D labs solve the matter by having inventors assign their patents to the corporation. Most other inventors tend to work in very closed circles, usually circles of one, glaring out at a conspiratorial world. What to do, then, in an open situation with no governing authority?

Fleet and I eddied off to the side to imagine a computer program that would do the job objectively. He floated the idea of tracing key words. I thought more safeguards would be needed. After all, how do you measure the value of an idea, especially if it's couched in a joke? Is there a value scale for jokes, other than a TV producer's spectrum of stony stares? Then, too, because a commercial on-line service must eventually trim its branches, it would be all but impossible to decide which offshoots should be kept with the original tree.

Conversely, if every offshoot were included, you could quickly end up with 30 million shareholders of unknown repute.

These problems all revolve around a central ambiguity: what kind of parameters could make an on-line conversation legally binding? How, in other words, to ensure a form of doing, rather than just some paltry form of saying?

This is a question well worth answering. For all the talk about the end of intellectual property rights, the bug-bag episode is one of the first serious attempts to address the idea of remuneration for on-line labor done. Will it work? Do I have any idea how big or small an idea it is? Of course not. But I'd be willing to bet it's bigger than a breadbox.

November 29 – December 5, 1995

Yankee Ingenuity Embattled

ON MY DESK, I have a map of the United States with clusters of marks inscribed in bright, formica orange. I decide where to place these marks by what you might call statistical eavesdropping. Whenever I hear someone complain about malls, condos or chain stores ruining the character of a region, I make a corresponding mark on the map. If I hear the same diatribe repeated by someone else, I add another mark to the first one. Some states are in no immediate danger of getting inked up—South Dakota, West Virginia—but most are suffering more or less continuous contact with my pen. One state turning orange particularly fast is Maine. And right about now, the front is sweeping through a once sleepy town known as Blue Hill.

I first went to Blue Hill before I knew how to speak or walk, and I've been going there almost every summer since then. When I was still in kneepants, I would walk to the general store a quarter of a mile away to buy Nehi soda and *Classics Illustrated.* The phone line at our house was shared by three other houses, each with its own distinctive ring. There was an active henhouse down the road. The center of town, some six miles away, had exactly one boring restaurant to its name.

Today, all that has changed, as tourists discover a Blue Hill that's already gone. Or rather, a Blue Hill that's *almost* already gone, because somewhere behind the satellite dishes and the lobster iconography, the local culture is still hanging on. And since Blue Hill is a stolid New England town, the term "local culture" is pretty much synonymous with mechanics and inventors. Basically, we're talking about one of the last pockets of Yankee know-how. Some far-flung cultures are trying to maintain their tribal dances. Blue Hill is trying to salvage its native tradition of tinkering.

The depth of this tradition was brought home to me this summer as I paddled up Eggemoggin Reach in a kayak. Six of us were on an afternoon tour, slicing through the summer waters, on the lookout for cormorants, maybe the occasional seal. Between huffs and puffs, I recalled that this was the very area where Buckminster Fuller grew up. I also recalled that when Fuller was a boy living on Bear Island, he rowed these very waters every day to pick up the mail, and that he was doing just that when he thought up his first invention. Instead of regular oars, he imagined a single oar, constructed something like a jellyfish. Push the oar off the stern and the boat surged forward. At the end of the stroke, the oar collapsed into itself. Bring it in, push again, collapse, bring it in.

While I marveled at this tone poem of an idea, our guide pointed out an impressive building on the nearby shoreline, a lighthouse design done up as a castle. "The Beamis house!" he shouted. "Built by the guy who invented the folded paper bag!" Serendipity rarely gets better than that.

Of course, it's been a long time since Fuller and Beamis dreamed large in these woods, but there are plenty of living inventors who have taken their places. No one who visits the Blue Hill Fair, for example, will fail to notice the bright grids erected directly across the road, at the headquarters of one Miles Maiden. Among other things, Maiden is building solar panels that maximize the intake of energy by following the sun in its path. Then there's Peter Zinn, inventor of the flat-panel keyboards seen in McDonald's franchises the world over. Not far away from Zinn's estate stands the house of Alfred Martin, whose father invented the fifth tractor wheel. Over in Bar Harbor, there's a guy whose specialty is weather devices. Spit anywhere in Blue Hill and you're likely to hit an inventor. In fact, the genius loci is strong enough to support a patent *illustrator.*

John Gallagher lives in East Blue Hill, some miles from tourist-clogged Blue Hill proper. To get to his house, I drove over the steel bridge and up the rise until I saw the windsock that marked his driveway. In his yard sat a pair of antique cars, looking gorgeous and unused. He saw me arrive and invited me into his rambling, clapboard house.

Gallagher made his first patent drawings in the mid-'60s, when he was still a teenager. By the '70s, he had moved to Maine and was working for Carl Booth, a patent illustrator based in Boston. Today, he's basically his own man, and he gets enough work from Boston attorneys to underwrite a second career as a painter. To hear him talk, it's a pretty swell gig.

"It could be anything really," he says, airlifting a cup of tea into the vicinity of my face. "I've had to draw GI Joes, laser-protector gloves, the Big Shot Polaroid, Speidel watchbands." Once he had to field the demands of a classic gonzo inventor, whose windmill defied all common sense in its complexity. "He had a bad case of inventoritis," sighs Gallagher, shaking his head.

It occurs to me that patent illustration is a fairly parochial activity, with its own esoteric rules and terminology, and Gallagher is quick to agree. "Completely esoteric," he says. Then, pulling out a pad of paper, he shows me some of the tricks of the tarde. A patent illustrator, he says, has to draw an object with the shadows to the bottom and the right. The curve of an object seen from the side is shown with a series of straight lines that gradually come closer together. A cross section takes a field of diagonal lines.

On its own, this visual shorthand is fairly simple stuff, but patent illustration has its own special set of demands. Care must be taken not to confuse the symbolic lines with structural ones—in the end, the whole shebang has to look like something that functions. It also has to get that way fast. Unlike architects or engineers, who labor with an eye to the long haul, Gallagher likes to turn an application around in 10 to 20 hours. And surprisingly, he almost always works without the aid of a computer.

"Except for schematics," he says, "which involve electrical circuitry and things like that, it's almost always easier to do it on paper."

So I guess that's the consolation prize: computers haven't reached into every last cranny of Western civilization, even if the cappuccino culture has. Out at the edge of the formica-orange mass, a soul can still get by with an eye and a hand for mechanical solutions. But as I walk back down the driveway and admire Gallagher's antique cars again, I wonder how long it can be before the patent illustrators and inventors of Blue Hill and environs succumb to the CAD/CAM wonder worlds, the programs that tell us how everything works, the master circuits that wire the brain into place. And I wonder too at the supreme irony that is the inventor's lot. After all, wasn't it the inventors themselves who designed the juggernaut that now threatens to crush them?

The malls roll over America; the fast-food chains are calling me. I throw my car into drive and rumble back into the fray.

September 25 – October 1, 1996

2

TRAINS, PLANES AND SPACE ELEVATORS

A Bicycle Built for Eight

"THAT AIN'T NO BIKE." The driver peers out his car window and develops his argument. "Unh-uh. That definitely ain't no bike."

Technically, the man knows whereof he speaks. Octos, a circular vehicle pedaled by eight riders—seven facing inward and one steering—is a quadricycle. But some beliefs die hard. At the next intersection, a helmeted man touches down on one leg and kneads the bars of his own, technically unassailable two-wheeler. "Five hundred pounds? That's too much, man. Mine here's only twenty-three."

The light turns green and we move on. Puzzled pedestrians part in a wave as we enter Central Park. Their eyes reflect the crispness of a blue September day, but their tongues wag with impetuous, I'm-gonna-wet-my-pants curiosity: What's it called? Where's it from? And first and foremost, what the hell is that thing?

And so a synopsis. Octos was designed and built by the husband and wife team of Eric and Deborah Staller. The eight sets of pedals connect to a universal joint within the circular tube, making a single-gear, direct-drive system. It's not for sale. There is only one existing version. The Stallers keep it at their home in rural Pennsylvania. It is slightly less than one lane wide. It can go forward, and it can go in reverse. It can attain a speed of about 20 miles per hour. The brake and steering mechanisms are accessible only to the driver. It really does exist.

"It's one of our urban UFOs," says Eric, gazing at me with his stainless-steel eyes during a break at the Boathouse Cafe, the lot of us dressed in black-and-white tights that lend us the air of outpatients undergoing mime therapy. "We think of them as metaphors for future travel. People have this fear of the future, that technology is going to be dehumanizing, or that the future will be dehumanized by technology. With our pieces, we try to say, 'Yes, the future is going to be more technological *and*—it *can* be human.'"

The Stallers have been creating their decidedly humanizing metaphors for some time now, beginning with one that's become pretty familiar around New York—a Volkswagen bug covered in thousands of tiny lights. Their more recent Bubblehead is a bicycle built for four riders, each of whom must sport an illuminated bubble on his head. Like many artifacts that demand to be seen, Bubblehead has made it into pictures, in this case, a cameo appearance in the Michael J. Fox movie *For Love or Money.*

But Octos is different from the other urban UFOs. "The earlier pieces used a lot of what I call gratuitous elements," says Eric. "I guess Octos is more of an invention. It's shaped by what it needs to do."

Tim Mills, the Stallers' engineer, toyed with the idea of getting a patent—until he learned how much they would cost, that is, at which time he abandoned the notion posthaste. (A patent can end up costing $5000 or more.) Not that the Stallers are all that worried about renegade replicas. Press the point and they will shrug and look a little bored. Indeed, the world of patents, with its vigilant lawyers and its infringement thieves, seems a little irrelevant when you're showing off your pride and joy before the minions of Manhattan.

Gifts to the universe notwithstanding, the Stallers can seek solace in the deterrents posed by production. The $40,000 price tag alone is bound to discourage knock-offs. And besides, as Deborah notes, making an Octos is just too much damn work. Who would bother to rip it off?

Well, funny she should ask, because while we're still lolling around at the Boathouse Cafe, we're informed that someone is "playing with your thing." Translation: someone is *stealing* your thing. Eight people, in fact.

Eric runs over to find an octet of citizens already pedaling way on his creation. A brief conversation ensues, during which these good folk have the audacity to stake themselves as the rightful owners. One of us asks one of them if he would steal a car in broad daylight. "Sure," he answers. "If the keys were in it." Then, just as quickly as the situation has developed, they disperse—in different directions.

It turns out they hadn't even known one another before their conspiracy gelled.

The mood that follows is a little tense, as might be expected. We continue along on the cool pave of the park, ad libbing ourselves back into a buoyant mood. A man on a recumbent bicycle—those low-slung vehicles that allow the driver to lean back and stretch his legs—talks shop with Deborah for a while.

"We're contemplating building a recumbent vehicle ourselves," she explains after the man has pedaled away. "If we do it, it will have six pairs of riders side by side. We're developing a wheel that can work on the street as well as on the tracks—it'll have a little bracket that holds on to the rails."

"There are a lot of smaller railroads in our area that are defunct," adds Eric from the driver's seat, his face turned in profile, "and the tracks are just sitting there, getting rusty."

The Stallers are also bandying about the prospect of converting Octos to three gears. Such a conversion would be quite a coup, given the complexities of designing a single-gear system, but Tim Mills is a can-do engineer, and he is already working with a gear box company to try to pull it off.

Our spirits revived with talk of future metaphors for future travel, we emerge from the park and begin our steady advance down Broadway to Washington Square. Cabbies tread asphalt beside us, holding up traffic and gouging the dandruff from their overworked heads. Distinguished men walk within millimeters of street signs. Sophia Loren never had it so good.

As we near Forty-second Street, a police van pulls up beside us. For once we feel as entitled as the law. They point at us. We point at them. There are eight of us in black and white, and eight of them in blue. Finally, the obvious question is posed: Can they get on?

Really?

Yes, really. So right there in Times Square, we pull over to a spot where no poor schlub has parked since goats grazed in Manhattan. A crowd immediately gathers. New York's Finest pile out, and damned if they don't climb aboard Octos just long enough for a photo op.

Which seems to suggest that, patent or no, cops and robbers will converge on a good thing all the same.

October 1, 1993

Sideways into the Future

ONCE UPON A TIME I wrote a pleasant little valentine to the blimp industry. No big deal. Just a glimpse into one of the more interesting subcultures (well, supercultures) around. But as I was soon to learn, my perspective then was Kansas gray to the color of Oz. Because that was before I had the chance to fly in one.

Let's put it right up front. You can forget about bungee jumping, mountain biking, S&M Scrabble. Flying in a blimp is, quite simply, a life-transforming experience. In fact, I was so completely blown away that, at the risk of spoiling the rarity of the thing (blimp rides are available to journalists, cameramen and government officials), I'm ready to argue my case for a blimp jubilee.

What strikes you right away when you meet an airship pilot is how much he loves his job. The first one we encountered at Teterboro Airport was a clean-cut man named Charlie. As he drove us out to the airfield, I asked him if he preferred to fly blimps or planes.

"Blimps," he replied without missing a beat. "No comparison."

Charlie then handed us over to the crew. We were flying in the MetLife craft known as *Snoopy One*, which seats four in a gondola roughly the size of a Toyota sedan. As we stepped in, we were greeted by Russell Adams, our pilot for the day. Right away, I could see the same endorphin-rich look in his eyes. "Can you imagine the dumb luck?" they seemed to be saying. "I fly a *blimp* for a living."

Passengers will find this look contagious. Liftoff, for one, resembles nothing more than an overgrown football being hiked. The crew members line up in formation—some on the wings holding tethers, others hugging close to the gondola. At a signal, two lawnmower-sized engines spring to life . . . and up you go.

Once you're aloft, there's a moment of shock, but this is mostly a reflex borrowed from heavier-than-air flights. In no time at all, you're floating over the Meadowlands at a cool 35 mph, realizing how much stress just isn't there at all. The sensation is distinctly amniotic. Gravity keeps the gondola on the underside, a cradle swaying under a benevolent helium teat. The *Snoopy One* is also nimble enough to manage some steep angles, allowing you to experience the unique phenomenon of facing the ground without speeding toward it. That's New York spread out before you, not getting any closer.

This kind of unharried view can spawn some unusual observations.

"I've noticed that New York has gotten a lot cleaner," Adams told us as he guided us over Manhattan. Four years ago, he explained, the topside of the city looked pretty ratty. He pointed out the new Chelsea Piers complex as an example. From street level, this building looks a lot like a warehouse with a new coat of paint. From the air, it's a truly impressive piece of architecture.

And so it went. We cruised over midtown and discovered how wafer-thin a skyscraper can be. We orbited the peaks of the twin towers, eyeballed the crowds.

Adams deadpanned: "Must be really scary up there."

We circled and circled. The wind blew clean. We lay in the palm of the sun.

IF FLYING IN A BLIMP is an experience granted only to a few, I certainly didn't mind being one of them. But a day or two later I got to wondering about it. After all, I reasoned, if satori can be so easily had, why not offer excursions for anyone willing to shell out the price of admission?

Then I remembered the march of progress.

It's difficult to overstate how completely the goad of progress has come to influence our thinking, despite its recent entry into the historical stream. No matter its shape or form, the new is almost always presumed to be better. This explains the outstanding benefits of the laser printer, but it also explains the head-splitting triumph of the car alarm. As a race, we seem compelled to rush to the cutting edge. Sometimes this gives us the edge. Sometimes we just get cut.

It doesn't have to be this way. In previous columns, I've mentioned a number of viable ideas that have been lost to the linear imperative, most notably DC electricity and the steam-powered car. When these technologies breathed their last in the 1930s, the reasons for their demise seemed sound enough. Today, however, the old reasons no longer hold water. The widespread use of DC power would go a long way toward decentralizing political power in this country. Steamers, meanwhile, could solve most every problem currently associated with automobiles.

Of course, not all extinctions are equal, and airships went out in a particularly vivid burst of dishonor. In fact, thanks to some film footage and a live radio broadcast, the 1937 crash of the *Hindenburg* put the kibosh on the zeppelin business altogether. Blimps, also known as non-rigid airships, emerged soon after that. At first, they were used mainly by the Coast Guard, and eventually a few commercial outfits followed suit, but despite the switch to helium and flexible envelopes, they never really caught on. Today, according to Adams, only some 15 blimps cruise the American blue.

Obviously, the memory of a hydrogen blaze has a lot to do with the shortage of blimp veterans today. But, really, how accurate is that memory? What was shocking in the '30s pales before the terrors of the '90s. Before stepping into the *Snoopy One*, we had to sign a waiver, and while that's fair enough, you really have to wonder why commercial airlines don't require the same. After all, there's nothing quite like the jolt of metal hitting the Earth at 500 miles an hour to make an insurance company fret.

You also have to consider the difference that media makes. Call to mind any commercial jetliner crash from recent years. How many of them were captured on videotape, much less on a live broadcast? With all due respect, if a photograph of the bodies pulled from TWA Flight 800 were published in a major newspaper, the jet would be history, too.

So we have learned to spin our grief, and the passenger blimp remains the best-kept secret of modern civilization. This is a shame. I'll be the first to admit that blimps will never answer all the demands of transportation. They aren't fast enough for cross-country flights, and they can't fly in thunderstorms. Still, they would serve quite well on puddle jumps—from New York to Washington, for example, or as a replacement for the Long Island jitney. They would also be ideal for recreational excursions. Think the Circle Line of the air.

As a matter of fact, if New York City Mayor Giuliani is really serious about his quality-of-life program, he could not do better than to institute a program of commercial blimp flights around New York. According to Adams, the *Snoopy One* cost about $1.5 million to build. Even when you add maintenance costs, that's not much to ask for a two-hour ride that could change the lives of countless New Yorkers. Especially when you consider the alternative. I mean, what kind of life do we really expect to get from an unexamined onslaught of the new? More car alarms? More crashes?

Sideways, I say. Sideways into the future.

August 26 – September 3, 1996

New Blimps for Real People

YOU'VE HEARD ME argue that ordinary people, as a rule, should be given the chance to fly in blimps. Now it looks as if my prayers might be answered. Recently a new breed of blimp left its mooring in Hillsboro, Oregon, for an undetermined West Coast address, where its owner is expected to give it a good coming-out party before the summer is over. This blimp is the first of the Lightship A-1 series, and while it will function perfectly well as a billboard, it was specifically designed with sightseeing tours in mind. Not only that, but two other companies already have dibs on their own A-1 blimps. Six decades after the infamous Hindenburg crash, the long moratorium on passenger airships may finally be coming to an end.

The driving force behind the A-1 series is Jim Thiele, the founder and CEO of the American Blimp Corporation. Indeed, Thiele is the driving force behind most of the familiar blimps in the sky today, and in many ways, his own career has been the story of the blimp's protracted rebirth.

As a boy in Topeka, Kansas, Thiele already exhibited every sign of being what insiders call a "helium head." He was still in high school in 1975 when he built his first hot air balloon, slapped an advertiser's name on it and flew it at a local shopping center. By the early 1980s, he had graduated from the University of Kansas with a degree in aerospace engineering and was beginning to work at airplane companies, including ILC Dover, the company that manufactures spacesuits for the space shuttle.

All during this time, Thiele had been aware that

blimps were plagued with problems, but he had always assumed that they would be fixed without his help. It took a disappointing stint at an airship company, from 1984 to 1986, to convince him otherwise.

"I founded the American Blimp Corporation in my living room in July 1987," he told me by phone last week. "I was in Eugene, Oregon, at that point. I was unemployed and single, and I guess I felt that losing everything I owned was a reasonable gamble, since I basically had nothing." By 1990, he had moved his company to Hillsboro, a suburb of Portland, and had raised enough capital to build the Lightship A-50, which, as he puts it, "contained every imaginable idea I had for improving airships."

The problem with blimps, as Thiele saw it, was their failure to deliver enough bang for their buck. Having been designed mostly for military use, they were poorly adapted to advertising. Worse, anyone who wanted to buy one had to spend upwards of $5 million.

His first step, then, was to produce a fabric for the balloon that was weather resistant, clean and, above all, translucent. The effect of this proprietary material, which gives the Lightship its name, will be familiar to anyone who has seen one of the MetLife blimps at night. Unlike older blimps, which are studded with electric lights, the Lightships are illuminated from the inside. This not only gives advertisers better coverage, but it does so at a lower cost and—always a critical factor in airships—a lighter weight.

Thiele's other big change involved the cables that connect the balloon to the gondola. The purpose of these cables, aside from keeping the blimp together, is to distribute the weight in the gondola—the payload—throughout the larger area of the balloon. In order to balance the payload well, a blimp needs dozens of cables, each "tuned" with a turnbuckle to a specific pressure. Traditionally, these cables have been strung between a curtain inside the balloon and the top of the gondola. What Thiele did was run his cables along the *outside* of the balloon and then attach them to the gondola floor.

These advances, along with some incremental improvements, led to ABC's debut production models: first the A-60, and eventually the larger, more widely adopted A-60+. The difference was dramatic. For one thing, their exterior-cable system cut the normal 30-day setup time (including shipping and inflating) down to five, which meant that Lightships could be shipped on short notice. As a result, clients who otherwise might have passed were suddenly in a position to place orders. (Since then, Lightships have flown the skies of every populated continent at one time or another, thanks in part to ABC's quick response.)

The A-60+ was also more efficient once it was up in the air. Where Goodyear's blimps needed 200,000 cubic feet of helium to carry seven passengers, the A-60+ needed only 70,000 cubic feet to carry five. And where traditional blimps called for as many as 24 crew members, the A-60+ demanded only 14: a pair of pilots, 2 mechanics and a ground crew of 10. Perhaps most important, Thiele and his engineers were able to bring the sales tag for a blimp down to under $2 million—less than half the going price. (The advertising rates are comparable. Prior to the Lightship, people were attempting to charge over $300,000 a month; today, an A-60+ typically goes for less than $200,000 a month.)

The Lightship A-60+ was widely hailed as a success and has since become an industry standard. In fact, it may be the ubiquity of the A-60+ that got people thinking about blimps as passenger vehicles. People like me, for example. In any event, Thiele began to see a demand for sightseeing blimps and set out to satisfy it.

The result was essentially a Lightship doubled in size. The blimps in the A-1 series—the A-130, the A-150 and the A-170—all have 10-seater gondolas, and their balloons hold roughly twice the amount of helium. (Multiply the series number by 1000 and you'll have the dimensions in cubic feet.) The A-150 is considered the flagship, but the other models have specific advantages to commend them. The A-170, for example, is well suited to high-density altitudes like that of Mexico City. (Density altitude is a term that combines the variables of elevation and temperature. In hot, mountainous regions, the density altitude will be especially high, forcing blimps to work harder.)

But what distinguishes the A-1s from the earlier Lightships—and from all other blimps, for that matter—are the improvements in acoustics. Until Thiele came along, airship companies simply screwed airplane propellers into their engines and let the devil take the hindmost. Unfortunately, airplane propellers are god-awful noisy and not at all what your average consumer has in mind for a peaceful respite. (Having flown beside them myself, I can vouch for that; imagine putting your ear up to a lawnmower.) So ABC hunkered down and designed a new propeller from scratch. To hear Thiele tell it, these efforts paid off handsomely. "The A-1," he says, "is the quietest blimp in the air today."

Of course, in making a larger balloon, Thiele has been forced to forgo some of the advantages he worked so hard to create. He'll be the first to admit that the A-1s run to the heavy side, and that they're fairly expensive as Lightships go. And so far, buyers seem to agree: the first three to sign on plan to use their A-150s for promotional purposes.

Still, as airships geared for the masses, the A-1s represent a significant step forward for blimp sightseeing industry. Not only are they designed to do the job, but they're actually in production, awaiting only the call of the magnate farsighted enough to see that what's good for the minions is good for him.

Thiele, for his part, foresees some guaranteed crowd-pleasers. "Certain areas of the world will be ideal," he says. "The Great Pyramids, the game reserves of Africa, the Great Barrier Reef. These will all be good sites."

If ever there were a time to add New York to that list, that time is now.

July 30 – August 5, 1997

Props of a Stunt Man

LIFE IMITATES ART, imitates life, imitates birds. I think. Or maybe it's art that imitates life, that imitates death transfigured. It's hard to keep track. With every passing day, we blink out from deeper armor. Or so you would think from talking to Eddie Paul.

I almost wrote a screenplay once about a guy very much like Paul. In this ill-fated story, a stunt man decides to become a "real" actor, with disastrous results. A lead part in *Beowulf* lands our man outside an abandoned Quonset hut, decked out in fur, rehearsing his lines. Meanwhile, the crew inside is falling in love with the mechanical Grendel they have built. Madcap capers and a terrible end await, of course.

Thankfully, Eddie Paul's life has not been like this. Not really. But it's been weirdly close. Paul started out designing hang gliders—"back when people were getting killed doing it," as he casually puts it. A couple of years of this and he was ready for the second layer of reality: he was asked to build the props for a movie called *The Wright Brothers.*

"We built three gliders," Paul explains when I catch him at home, "and I went up to deliver them, and then they wanted me to teach the stunt men how to fly them. They didn't know how to fly. They finally realized it was easier for me to fly them than to teach the stunt men."

Now, this movie was not just about the Wright brothers, mind you, who had a fairly good record where crashes were concerned. It was also about all those guys *before* the Wright brothers, who spread their various wings and met their singular dooms.

The most famous fatal birdman of the early aviation era was Otto Lilienthal, a veritable Icarus of a man. Lilienthal was technically the first human to fly, if you count gliding as flying. He was certainly the first to make it look great. Consciously imitating bird flight (that imitation thing again), Lilienthal stood on a lot of cliffs with his full span splayed and a pair of goggles that made for some dashing photographs. The bad news was that he crashed one day. Birds don't often do that.

"He crashed when the keel came forward," says Paul. "Basically, he died from a broken back."

I tactfully offer that Otto Lilienthal's brother, Gustav, told a different version, but there's no point in pressing the issue. It was Paul, remember, who had to go up in his reconstructed Lilienthal machine and live to tell the tale.

"I redesigned the keel," he continues now. "Otherwise it was identical to his. I put a plate in so the keel couldn't come forward. Wouldn't want to make the same mistake."

That was how it began. History notwithstanding, Paul survived his altered imitation of an imitation of a bird, got his A-card, as they say in the stunt biz, and started doing work for *The Dukes of Hazzard.* Before long, he was building his own machinery for stunts, which meant that, unlike my besotted character, Paul was controlling the monster, rather than the other way around. This is always handy when you hang off tall buildings and drive cars into walls for a living. "Unlike other stunt men," he muses, "who always had to try something and wonder if it would hold, I knew it would work, because I built it. From the other side, I not only got to invent things but I got to test them out, too."

The American people will certainly be familiar with this portion of Paul's handiwork. If you've been wondering who invented the cars that drive up against a ramp and then turn over—a special piece of movie gadgetry called the "pipe ramp"—it was Paul, among others. ("A few people did it at the same time," he says, "but I was certainly one of the first.") From there, he did all the cars in *Grease*, built one of the first robot-controlled cars, coordinated the stunts for Peter Bogdanovich's *Mask*, you name it.

The driving thing probably peaked with Paul's "no-hands" motorcycle drive from Los Angeles to Las Vegas—a distance of 286 miles. To achieve this feat, which was not a movie stunt but a bid for the *Guinness Book of Records*, Paul remounted the throttle on the carburetor and then simply leaned when he wanted to turn. But that wasn't the hard part. The hard part was refueling. For this, he had to have a truck pull up beside him and pass a fuel line to the driver on something like a fishing pole. He refueled three times during his trip, leaned to and fro, and presumably took one hell of a piss when he was done. And after all that, *Guinness* didn't even accept his performance as the record.

"It was kinda stupid, I guess," Paul laughs.

Having conquered air and land, Paul progressed to the davy deep. For *Rescue 911*, he built a barracuda that jumped out of the water. For Jacques Cousteau, he built a mechanical great white shark, which actually swam, and a shark suit, which proved that sharks couldn't bite through a certain type of plastic. His mechanical diver was able to show that sharks don't like you as much if you're moving. They like a passive meal, just like you or me.

Of course, a stunt man will get bored easily, especially with sharks that smile and move on, so Paul eventually went for the big kahuna and tried to direct a movie of his own. Talk about terror; that cured him but good. "Let's just say that I didn't care for working with people who weren't totally honest," he says, choosing his words carefully.

Paul's deus ex machina was, fittingly, an invention. Called the CEM, for Cylindrical Engine Module, this bit of technology can act as a pump, a compressor or an engine, depending on your pleasure. Right now, one 30-pound model is putting out 110 gallons a minute for a firefighting outfit. Elsewhere, there's a buyer who wants to use it in a steam-powered car. (Incidentally, I once reported that Paul was stirring up a dialogue about workable steam-powered cars. This was not exactly right, but it wasn't exactly wrong, either. While it was someone else who actually brought up the Doble cars of the 1920s, it was Paul who answered with news about a steam car that may use his pump.)

"I think they're going off on the wrong angle with the electric car," he tells me. "I think it's going to be either natural gas, hydrogen or steam. It's a lot more practical than what we're doing now."

At this point, the CEM technology is doing well enough that Paul can support himself as a full-time inventor. He also started up a think tank, called I-Cubed, which will assess patents for wayward inventors. And he's working on a kind of harpoon that will shoot from the front of a cop car to stop a car full of perps. And, and, and . . .

Yet the imitation thing keeps dogging him. Still on the phone, I look down and notice from a fax he sent me that he lives right by the L.A. airport.

"I imagine your shop looking something like a hangar," I suggest.

"Actually, it's a Quonset hut," he replies. "One of the few left from the '40s."

A Quonset hut? Goddamn. Maybe I'll finish that screenplay after all.

November 1–7, 1995

Gravity and How to Use It

LAST WEEK, pedestrians passing through Astor Place had their dense interior monologues interrupted by the sight of a large blue vehicle parked in the shade of Cooper Union. Interrupted, that is, but not necessarily explained.

When I came upon this conspicuous object, known as the Gravity Kinetic Transfer Transit Circuit—or Gravtran for short—no one seemed to be minding it, so I stepped in. There was just as little exposition waiting for me inside as out. Except for a few documents taped to the walls—inscrutable screeds pertaining to legal disclosures and engineering concerns—the car was entirely empty. Just me and a big tin can, then . . . until a couple came in and started panting in my vicinity.

"Everything I've done so far," said the rangy but handsome young beau, cupping his darling's face in his hands, "has been leading up to this. I'm not who I said I was."

Indeed he was not, as he soon proved by skipping off with his paramour in a flurry of laughter, leaving me and my notepad behind.

Suitably distracted, I stepped outside and started taking down the sundry information posted on the outside of the car. One-hundred percent fuel-free . . . zero pollution . . . an 800 number written on paper and affixed to the bow with masking tape . . . The vehicle itself was hitched to a Chevy pickup, in back of which, sticking out from a shapeless tarp, lay a few plastic dolls. All this was interesting enough, but it wasn't much without a live body to decipher it, so I sat back and waited. Finally, a gruff man in a cowboy hat and salty white beard arrived, opened the door to the pickup truck and got in. When I knocked on the window, he rolled it down.

"You're the first person from the press to tell me who he was beforehand," he said, giving his name as Sam Dordoni. I offered to buy him a cup of coffee, which he dispatched so fast I never saw where it went. But that was only par for the course, I guess, because he was just as adept at dispensing the facts on his patent-pending invention.

"Every piece of this vehicle," he said, "is made from recycled material. That round piece up there came off a truck cab—you know those curved things? The floor came from an Amtrak car. The sides are made from airplane wings. The wheels came from a crane. I've been here for fourteen days. Ran out of gas trying to keep warm in my truck."

With that, he unveiled a video monitor on the back of his pickup and promptly ran a demo on whatever was left of his battery.

To understand the secret of the Gravtran, it helps to imagine a set of tracks that begins at a height—say, 20 feet—and gradually tapers to ground level. Arrange these tracks in a circle, set a car on the tracks and the only obstacle to an endless round-trip ride will be a method for lifting the car to the starting position. Of course, most engineers faced with this dilemma will simply install an engine and some kind of ratchet and call it a day. Dordoni, however, has come up with a cannier solution.

When passengers get into the Gravtran car, their own weight acts on a pulley, in turn raising a heavy weight into the air. This weight is then locked into its raised position while the car is set free to trundle down the tracks by force of gravity. When the car reaches the end of the track, the pulley system is activated again and the original process is reversed: as the passengers get off, the weight drops down and the car is raised back to the starting position. No sleight of hand. No jury-rigged theories of physics. Just a short-range transit system requiring no fuel at all and only a few adjustments for friction and speed control.

"It's a trick is what it is," says Dordoni. "But it'll work for forty, fifty years without any maintenance. And people like it—the students here at Cooper Union especially. They even made up a petition and got people to sign it. Look at all these names."

It's not hard to fathom why these particular students would be so keen on the Gravtran. After all, their alma mater was founded by the same man who built one of the first American locomotives from scraps lying around his backyard. Unfortunately, their enthusiasm has been less than completely effective in places. Before Dordoni arrived, I noticed a parking ticket fluttering on his windshield. And in a more direct assault, someone has used his petition to pen in the verdict of "Whacko!"

"Look at that," mutters Dordoni. "The only one out of five hundred people who said something negative. Bastard. A person would have to be mentally poor to write something like that. I'm out here trying to do something for the people and that's all he can say? I'm telling you, things are getting much worse. I was in the Army in the sixties and that was all supposedly for nothing, but today it's much worse. Today, these kids go in *knowing* that it's all for oil. They know it when they go in!"

This sore point out of the way, we proceed to the interior of the vehicle. Dordoni points to a few diagrams, gives a few dates, describes how the Gravtran would alleviate the congestion at Giants Stadium. Then he closes the philosophical loop.

"If you think about it," he says, "the transportation systems we use now are mostly fighting gravity. The most important thing about a plane is for it not to go down, right? Same with a boat: if it goes down, you're in trouble. Even a car has to work against its own weight. So basically, we're fighting wars to fight gravity. All I'm saying is, if you can work with gravity, why not do it? My system won't work for everything. It won't replace every form of transportation in existence, but for certain things it will work very well, and very cheaply at that."

All this time, people have been poking their heads into the car, pausing long enough to wrinkle their brows or express their admiration. Dordoni pays them no mind until a tall black man appears at the door, at which point his soliloquy comes to a sudden halt.

"Johnny Rainbow!" he shouts. "You should put this man in your article. He's helped me out a lot. A *lot.*"

Rainbow—who in derby and tie radiates a kind of preternatural charm—introduces himself as the man responsible for getting Duke Ellington on both a postage stamp and a Manhattan street sign. Beaming at me from the distance of a different era, he hands me a flier announcing his upcoming engagement at Danny's Skylight Room: "The internationally acclaimed John Rainbow makes a rare New York appearance as he brings his uniquely expressive baritone and 'intimacy' into every note of every song." I have the distinct impression that he bows before leaving, although this can't be right.

Dordoni, meanwhile, has had his stream of consciousness diverted into more personal matters. "My father was a musician," he says. "He used to play in the Empire State Building, in the Garden State Jamboree. I used to go up there to hear him play and it was always such a good feeling. But then we would go outside and suddenly there were cars everywhere and lights and confusion, and the feeling would go away. That made an impression on me, you know? I guess I started looking for ways to keep that good feeling from going away."

It would be hard to find a better reason for inventing anything, be it a new kind of car or a better toothbrush. And if the occasional appearance of lovebirds and showmen is any indication, Dordoni is doing pretty well at that.

December 17–23, 1997

Utopia in a Briefcase

A CO-WORKER LEANED her head in my doorway: "Jack Marchand."

"Great," I replied. "What line?"

"No. He's not on the phone. He's here."

A flare went up in my mind. So far, I had only sent Jack Marchand a letter asking him to call me at my work number. I hadn't even spoken with him, much less made an appointment, yet here he was in the flesh. My meetings with inventors tend to be awkward at best. In the offices of the coffee table magazine that keeps me in the pink, it was bound to be a geek show.

Marchand apparently felt the same way. He came into my office furtively, as if not only to protect his elderly frame but his vast inner life as well. He glanced around the room, his eyes in deep retreat.

He would be in the Wall Street area that evening, he said. Could we meet at the McDonald's next to the Roy Rogers below Fulton Street? From his overpopulated briefcase, he then produced a piece of paper blackened with drawings, commentary, suggestions, maps and triply xeroxed fragments of patents. On the top of this samizdat production, larger letters spelled out "The Street Rag," and underneath, "What's Also Fit . . . But . . . Not Print."

In fact, I'd been studying another issue of *The Street Rag* at home. Having received it through three degrees of separation, I'd been puzzling over meticulously rendered diagrams for stadium rods. I'd been trying to piece together scrawled descriptions of an electric car. And I'd been teased by the many truncated remarks:

"Back copies are available. See distributor or vendor. Also ask friends and others for them."

See distributor? Ask friends?

That night, I arrived at the Roy Rogers near Fulton Street as agreed. I didn't see a McDonald's anywhere. It occurred to me that Marchand had the kind of face that dissolves in your memory, making recognition an exercise. His briefcase, on the other hand, was distinct enough. I recognized it right away as he approached me on the busy Broadway sidewalk.

After exchanging tentative greetings, we decided to go to the Roy Rogers, with its hound dog–blue Muzak and a handful of shills for customers.

To read his resume, Marchand has held 34 jobs involving some kind of invention work, most of them for things like hydraulic reciprocating grind head units and pressure gauge control units, although he claims to have designed an early CAT scan machine for Columbia Presbyterian Medical Center. He has been awarded at least two patents (#3,381,446 and #3,851,441). But all that, it seems, was before. These days, his sights are trained on more exotic game, ideas that veer closer to poetry—satellite systems, user-owned utilities, global highways.

Why his imagination has followed such an arc is not easy to know. My probes into his personal life got me nowhere. I knew he received mail, but whether he actually lived at the address was a question he managed to deflect. He never did tell me who his "distributors" and "vendors" were, or where they might be found. On the other hand, his inventions seemed to consume him completely, burning away such detritus as despair, or even passion, in the process. His psyche *was* his ideas.

Clearing the crumbs from our table, I asked Marchand to talk about his three-phase proposal for electric transportation—not a difficult task, since he has been developing this idea for nigh on 30 years. He even showed it to the government back in the '60s. For a while, he said, the Department of Transportation took him seriously (or at least pretended to), only to recede, as behemoth parasites will, behind the usual silent shroud.

The building block of the idea, the irreducible unit, is a replaceable battery. Cars would have two of them—one in the front, one in the rear. Trucks would have a side bay, possibly, to accommodate many batteries. Mechanically, it would function much like an oversized computer disk. It could be popped in and popped out on command.

As for the replaceability aspect, vehicles could pull up to a refueling station whenever the batteries began to run low. Marchand imagines these stations as part toll booth, part assembly line. A conveyor belt would draw your car sideways, ejecting the old batteries and inserting new ones in a three-step system. The whole thing could take place right on the road, pit stop–style. In the meantime, the drained batteries would be recharged inside the station itself, traveling up and down a number of columns in an automated recycling process.

"You're in and out of the station in one minute," Marchand said, polishing off his cinnamon raisin Danish. "Buses and trucks take two minutes."

It's a concise format on the face of it, and I can see how all those years spent on pneumo-digestive gristlehead shifters led to a broader application. But I was not to linger over the joys of industrial thinking made beautiful. "The next phase is the electromagnetic propulsion system," Marchand said. "I call it Mach-lev. Not mag-lev, which is short for magnetic levitation, but *Mach*-lev."

And with that, he unfurled a narrow strip of paper that extended well beyond two table surfaces. On it, he had carefully drawn an enclosed tubular highway with endpoints designated "Boston" and "Chicago."

"Once this becomes developed," he said, hunching over and pointing with his pen, "you're only limited in speed by how well you can align these tubes. It becomes an electron gun."

"An *electron gun*?"

"Oh yes. It's very feasible with this system that people could travel at two thousand miles an hour."

As I looked around the interior of the Roy Rogers, the hitching-post partitions and plastiform tables became visible to me at the speed of light. So, for that matter, did the peculiar skin-like texture of the napkin in my hand. Dispelling the image of electrons tooling down I-95 would take longer.

"Excuse me?" I ask, trying to hear above the onslaught of canned Christmas carols. "Did you say your electric car will go two *thousand* miles an hour?"

"There's no reason why not," Marchand replies, looking for all the world like the Fuller Brush man with a surprise pitch to make. "If you make a vacuum in these tubes, there's no resistance anymore."

The highway of which he speaks would be a one-lane tube that forms an immense loop, with terminals at major hubs. Once an electric car got on, he says, the electricity in the tube would take over, determining the speed of travel. Mach-lev would not sustain speeds anywhere near the speed of sound in its first incarnation, although 500 mph would be feasible.

However you slice it, this is one fast vehicle. And as it turns out, Mach-lev also solves a problem familiar to anyone who has tried to get an electric car on the road—the difficulty in making a car that can drive long distances without running the battery down. Still, you could take a wild guess that the government is a long way from accepting such a proposal, what with the expense and all. Marchand answers this caveat with the claim that his highway would be 10 times cheaper to build than its conventional three-lane counterpart—and 1000 times cheaper to maintain, because there would be much less wear and tear. The consumer, he insists, would also feel less of a sting. At 10 cents a minute, the 95-minute shot from New York to Chicago would cost a mere $9.50.

But Marchand doesn't stop here. We're talking about a single standard system for the entire world. Phase Three of his plan would see the construction of an international highway network stretching from Lesotho to Buenos Aires by way of a tunnel beneath the Bering Straits. Anyone with a hankering to see the Tokyo skyline could be there later today. True, this phase would need a $300 trillion investment over the next three decades to make it a reality, but on the scale of worldwide government budgets, that isn't as big as it sounds.

If you're wondering just where the juice for this high-tech Oroboros is supposed to come from, Marchand is one step ahead of you there, too. He envisions a worldwide solar energy network, with panels set up in the major deserts and large conduits attaching them all. Because of the placement of these panels, the sun would never set on the energy empire. These utilities would also be the property of the species, with a nominal salary to be doled out to maintenance crews. "They would get their pensions and what have you," he explains, "but it would be owned by everyone."

Just how we move from the concentration of wealth to a global stockholder system Marchand doesn't say, but at least no one can accuse him of thinking small. And as far as I can tell, his system does offer some compelling features: a sensible protocol for local travel, an affordable car ($3000), a vision of a sustainable economy worthy of Frank Capra and, lest we forget the virtues of pleasure, some truly exhilarating straightaways.

Indeed, the value here seems to be in the details, of which there are many. Motivated readers will want to check out Issue #16 of *The Street Rag*, which gives the most concise explanation of this electric Eden, along with a pithy editorial that lets the inventor's otherwise masked emotions be known: "Don't let the gangsters, our reps, our legal systems and other leeches lull you to sleep and then screw you and the world of what should be yours and available to all in the future at minimal usage cost, maybe even free."

Strong words from a guy who wears a tie.

Of course, Marchand is not the first dreamer to be whipped into bitterness after pitching an electric car project. Ever since 1894, there have been tentative unveilings of electric cars—and plenty of people only too willing to throw the veil back on. General Motors, for one, has played a part in the mess, helping destroy 100 electric railway systems in 45 cities between 1932 and 1956, according to a 1974 Senate antitrust hearing.

But lately the zeitgeist seems to be shifting. The Big Three have been mumbling about electric cars of their own for a couple of years, and now that Bill Clinton has promised to subsidize their race for the car of the 21st century, lo and behold, the prototypes are coming out of the woodwork. Chrysler is suddenly chiming about flywheel technology, and an upcoming issue of *Popular Mechanics* promises to show us "the world's first practical electric car," manufactured by none other than . . . General Motors.

All of which puts Marchand in the curious position of treading water in a new part of the ocean. After all, his inventions address more than just the type of energy used. They form a vast equation intended to produce a specific result—the people's car—and Hitler notwithstanding, the profit-minded have generally found such a product, well, *distasteful.* As a result, Marchand has adopted a novel tactic. Rather than spending time and money to secure patents, he is taking to the streets and buttonholing anyone who shows so much as a passing interest.

"It becomes an advantage for them to be a witness to the ideas," he explains. "Say somebody tries to steal one of my ideas. They file for a patent and get it. But now everyone knows it's not his idea, so they say the hell with it and they start making the thing, too. The guy who got the patent, he'll get challenged by the others—especially if they're going to make billions out of it. And eventually, it will come out whose idea it is."

In other words, sic 'em on each other and wait for your day in court. Not a bad gambit, either, as long as the first-to-invent system of patents obtains. (First-to-invent gives priority to the inventor no matter who files for a patent first.) But if first-to-file is adopted—as a host of big businesses are currently recommending, under the auspices of the World Intellectual Property Organization—Marchand's house of cards will collapse in a silent flurry.

At least he won't be alone in his sorrows. Should the patent system change, the overwhelming majority of inventors in this country—the same class of people who brought you the American century—could easily find themselves in Jack Marchand's shoes, peddling their ideas quite literally from a suitcase.

It's a sobering picture. Marchand has spent a lifetime trying to gain an audience with the government, only to find himself in a dismal fast food chain, trying to convey 30 years of intense bicameral thinking through a haze of bad confectioner's sugar. This is not the

Waldorf Astoria. This is not the way it was supposed to be. The script did not call for an employee wearing plastic gloves to come over and tell us it's time to go.

And so it's only par for the course, I guess, that the Muzak provides a maudlin Greek chorus for our exit: *"Counting the cars on the New Jersey Turnpike . . ."*

January 5–11 & 12–18, 1994
(Two Parts)

Solar Power in Space

WHEN THE 20TH CENTURY was banging on all cylinders, the great historians wrote histories of the future. Arthur C. Clarke, Isaac Asimov, Gene Roddenberry—these writers told us how it was going to be, and technologically at least, it all seemed pretty compelling. In a sense you could say their fantasies proved stronger than science. What's amazing today is not the sci-fi legacy of warp speeds, mindmelds and cosmic surfing, all of which has become part and parcel of the even dullest conversations. It's the fact that none of it ever really happened.

Gregory Matloff, a bushy-bearded, bespectacled man who exudes an unflaggingly chipper intelligence, doesn't seem to mind living in a past known as the 21st century. Co-author of *The Starflight Handbook* (1989, John Wiley & Sons) and erstwhile consultant to NASA, he gives the impression of loving his job to pieces. You won't catch him slinging mud or casting aspersions, even when baited. He's enjoying the stream of ideas too much.

"I don't know about you," he tells me over coffee, "but I love a dare. Particularly when people have said it's impossible."

The dare in question was posed in 1977, when his mentor, Michael Mautner, formerly a professor of biochemistry at Rockefeller University and more recently with the National Bureau of Standards, challenged him to come up with the theoretical basis for a solar-powered interstellar spaceship. That's interstellar, mind you, not some probe noodling around any old cut-rate planet. Thinking it couldn't be done, Matloff cited the experts who said as much.

Mautner wasn't fazed, however. "You just wrote a paper on ion drive interstellar ships," he answered. "*You're* the expert now."

Thus Matloff took up the gauntlet and became the first to work out the math for the kind of ship Arthur C. Clarke could only create with words. So large it would have to be built in outer space, the sail for this ship would be about one millionth of an inch thick, yet strong enough to withstand the heat of the sun. The word *gossamer* does come to mind. And if the ship could pass close enough to the sun (in order to get an honest-to-God head of steam worked up), humans could hope to reach the nearest star in 300 years.

The only problem is, humans just can't withstand an acceleration of 2000 miles per second. So Matloff scaled down to a more palatable level and did the figures on an interstellar colony accelerating at a much slower rate.

"You would only have to do this for a couple of hours," he explains, "and you could put the people to sleep during the approach to the sun. A nice joke would be to call it the Valium Trajectory."

Having blazed this trail for solar-powered transport, Matloff went on to formalize the mathematics for a generic model: a flat-sheet sail traveling on a parabolic trajectory at escape velocity (meaning escape from the solar system). And because it was generic, engineers could work on a prototype without the aid of outrageously sophisticated computers. The stage was set for bona fide hardware.

Enter Claudio Maccone, another challenger in Matloff's life. Maccone, an Italian mathematician, observed that this flat-sheet sail needed only a scientific application for development to begin. So they brainstormed, looking for the egg that might give birth to the chicken. "I was sort of embarrassed," Matloff admits, "because the answer was right in *The Starflight Handbook*."

The game show prize–winning category, it turned out, was not Travel but Communications. They calculated as follows: if you sent an interstellar vehicle out to a radius of 550 times the distance between the sun and the earth, the instruments aboard the vehicle would then amplify any signal coming from the opposite side of the sun by a factor of 100 million times. In other words, the sun would act as an *amplifier*, or a gravity lens. As a result, the solar-sail device could send and receive signals across a much greater range than is presently possible. And because the vehicle could weigh no more than 20 to 100 pounds, the feat could be achieved in a single human lifespan. Suddenly, a project conducted beyond the confines of the solar system began to look—be still, their hearts—practical.

"Astrophysicists love this," Matloff says, "because they would love to study the suspected black hole at the center of the Milky Way." Of course, the possibility of talking to someone out there was also pretty tantalizing.

So far, the project has been received enthusiastically. What started as a paper in the *Journal of the British Interplanetary Society* in 1992 soon became a $50 million project and has now blossomed into a $500 million flower. And with the Russian contingent looking the idea over, even more money may be on the way.

No doubt Matloff will never see a patent for the biggest ham radio ever devised. Even so, it would be hard not to cast him as an inventor. "What I do relates to invention in an interesting way," he says, "because I come up with ideas to ultimately break free of the solar system. We seem to be moving in the direction of a planetary consciousness, and it seems that we are in the nervous system of Gaia. Really what this means is that Gaia can reproduce itself. Gaia can outlive the sun."

Heady stuff, to be sure, especially if you think for two toots about a nervous system capable of reproducing itself onto another body. Yet for all his exuberant mind-stretching, Matloff still represents the inner reach of space technology. For a taste of the truly outrageous, you must turn to the work of one of his colleagues, an Englishman named Paul Birch. Matloff may want to go places in the galaxy, or at least converse with the inhabitants there, but Birch wants to change the way the galaxy looks.

Having heard of Birch through *Cryonics* magazine—another story altogether—I wrote to the *Journal of the British Interplanetary Society*, requesting any issues in which he had been published. A few weeks later, I received a handsome specimen containing two papers, the first titled, "How to Spin a Planet" and the second, without so much as a pause for effect, "How to Move a Planet."

Far be it from me to explain the technology required to do this, so let me hold fast to motivations and some simple blocking. To inhabit another planet, the argument goes, you first need to create earth-like conditions. Such an activity is known as terraforming. Now, Birch has his own particular views about this subject. Venus, he says, spins too slowly for earth life to thrive. But by exerting a kind of mechanical resonance, we could speed up its rotation in a matter of 30 years. Then, for really good value, we could widen its orbit by pulling it a few million miles farther away from the sun.

To be honest, this second proposal I found a bit disappointing. From a title like "How to Move a Planet," I had expected globes unhinged and amok in space, a little thrash-metal of the spheres. But alas, it seems terraforming is less slamdance than horticulture.

I suppose I should have guessed as much. After all, when the 20th century went kaput, the Americans started writing cyberpunk novels and shooting each other for laughs. But the British? Well, you know. They always were sort of weird about their gardens.

January 19 – 25, 1994

Planets, Rebuilt and Relocated

YOU MAY RECALL a column I wrote a while back that touched on the activities of one Paul Birch. This was the guy who wanted to make planets spin differently, pull planets into different orbits, stuff like that. Well, at long last, someone in the British Post woke up and pushed a bundle through the little slot—or whatever it is they push their bundles through—thus putting me in possession of what might be called the Expanded Birch Collection.

Most of Birch's articles appear in the appealing-sounding *Journal of the British Interplanetary Society.* Never mind that they could dispense with the word "interplanetary" without changing the gist of their title. This is not a sci-fi 'zine. It's a heavily legitimized periodical for thinkers who happen to be, well, way out there.

Birch himself clocks in with great frequency, no doubt because he's got such a rich inner life. If he's not talking about faster-than-light travel, he's waxing eloquent on genetic mutation, by way of a discussion on monetary inflation. In "Future Perfect?" an article that seems to be self-published, he slums a little and describes how to make a river that runs uphill. The task involves the creation of several meandering figure-eights, with sluices that cut across the loops. Draw as I might the diagram he describes, even I, a charter member of the Spatial Thinkers Pyramid Scheme, cannot master this contrary current.

More accessible, for all the math that comes in tow, are his ideas about space elevators. These massively underpublicized devices would provide a means for delivering payload to satellites. Sounds simple enough, but Birch wants to do it without the use of rockets. "As can be seen in Fig. 1," he writes, "a very long cable is needed, which must be able to support both its own weight and the weight of the space elevator." In other words, you could attach a giant dumbwaiter to your brand new geostationary satellite, and then you could send things up and down, if only you could keep the cable aloft.

Believe it or not, this topic is big enough to contain differing opinions. A certain Mr. Y. N. Artsutanov, for example, has apparently proposed the use of counterweights attached to the far end of the cable, with the satellite situated in the middle, like a bead on a necklace. The counterweight, says Artsutanov, would stay in place because of centrifugal force. As the earth rotated, the motion would throw the counterweight forever toward the void.

Not so fast, replies Birch. A cable long enough to accommodate these counterweights would have to be much stronger than anyone could ever afford. Better to stick with plans for low orbits.

What Birch has in mind is pretty complicated on paper, but after the thetas and deltas, it involves three elements: an orbital ring, a skyhook and a Jacob's ladder. The orbital ring—a kind of galactic hula hoop—would encircle the earth in low orbit. How it would stay up I don't exactly know, but if Birch says it will, I believe him. The skyhooks would then be attached to this humongous hula hoop, and the cables, now called Jacob's ladders, would be connected in turn to the skyhooks.

You will expect that Birch has anticipated the next step, and he has. Skipping over his hypotheses about extraterrestrial life, population explosions and sundry, we come to his discussion of supramundane planets. "Supramundane" translates not only as "above the earth," but also as "above the mundane," and nothing could be more precise. The idea is to build up one of these orbital rings until it becomes an artificial globe surrounding the mother planet.

"The underplanet or underbody is the underlying planet or other heavenly body," he writes, "which generates the supramundane planet's gravity." Lest we be denied a touch of the gallows humor he adds: "The apt term underworld may be less suitable for technical use."

Birch builds his argument slowly but surely, beginning with a method of transforming his orbital ring into a giant solar windmill, then moving on to the possibilities of a spherical mesh. He proposes lopsided orbital rings, planets that bulge in the middle, systems for constructing mountains all over your penthouse planet, you name it. He even goes so far as to suggest lassoing one of these orbital rings around the sun and using it for interplanetary travel. Needless to say, his ideas are ingenious. Why, the heavens are a veritable Tinker Toy set in his hands.

But forget about that. Since it happens to be nighttime as I write, let's take a look at the darker implications of this plan. The biggest reason for connecting a Jacob's ladder to an orbital ring, you see, is so you can send stuff up more cheaply, thus making it easier to build your own supramundane planet. Not around Jupiter. Who wants to go there? No, regardless of what his intentions may be, Birch is giving away the secret of how to build a new shell around Earth. And when the time comes, up the overlords will go, their latest stooges stationed at the moorings below, attended by the usual accoutrements: ramparts, SWAT teams, hired celebrities waving from the gate.

Ridiculous? Cheesy Asimov thrice digested? Perhaps. Yet Birch's main argument for the orbital ring scheme, you will recall, is an economic one. And it turns out he's not even talking about next century. He's talking about today: "A Jacob's ladder is much shorter than a cable to geosynchronous orbit would be, and thus does not have to be made of so strong a material. It is within reach of present-day technology."

Now, if the CIA once invented an exploding cigar for Fidel Castro—and if the young silicon turks are serious about their proposed new Disney World of satellites—then somebody must be lending half an ear to Birch's brainstorms. And imagine how the deal would go down. Busy Executive calls in from his car phone. "Gimme the word on this Birch asshole . . . what's that? . . . Right. Cables down to earth. How much? . . . Unh-hunh . . . That's all? Listen, Frank, I want a complete report on who controls the territory . . . That's right, who controls the [static] . . . What? What? Do I what? God damn this phone . . ."

Indeed, a little pre-emptive thinking casts wondrous new light on current events. Since Birch's orbital rings are based on the principles of geostationary satellites, they will have to be stationed above the Equator, and by extension, their ladders will have to be tethered down at zero latitude, too. Funny thing is, the Equator is exactly where you will find the majority of, that's right, the world's rainforests.

Well, well. Sorta makes you wonder what Mobil Oil and the other rainforest ransackers have up their sleeves, doesn't it? And, for that matter, just how hard the satellite pack are gnashing their teeth as they recite the time-tested rule: location, location, location.

April 13-19, 1994

3

ENERGETIC ENDEAVORS

Perpetual Motion from the Messiah

I HAVE FINISHED *doing monkey shows, and the energy technology will go into production on August 22 . . .*

So goes the message on Joseph Newman's answering machine, and I for one can find no cause to blame him. I mean, if you had invented a perpetual motion device, developed a unified field theory *and* discovered that you were the Messiah, wouldn't you be a little miffed by the naysayers of the world? Wouldn't a trivial detail like a rejection from the Patent Office confirm for all time the monkeyshine of man?

The message continues . . . *The energy technology is a gift from God and no power on earth can stop it . . .*

Truth be told, some of the powers on earth, far from trying to stop him, seem to have been helping him along. His spiritual saga began on D-Day, for example. While American troops waded to their slaughter, Joseph's older brother performed the smaller sacrifice of drowning at home. And while American troops secured the shores of Normandy, a halo appeared around the sun in Mobile, Alabama, giving the seven-year-old orphan Joseph his first sign that he was The One, though he didn't understand it as such at the time.

"A change came over me," Newman explained when I called him at his Lucedale, Mississippi home. "I changed to total goodness."

As he came of age, Newman's transfiguration proceeded apace. In 1954, a tour of duty took him to Puerto Rico, where the suffering of the children in the streets so enraged him that he sat on a mountaintop and vowed to do good even if there was no God. Upon returning home in '58, he began to get ideas about the makeup of matter. Thus the life of the savior became paired with the life of the inventor. At some point along the way, he found the wherewithal to invent plastic-covered barbells, which earned him a mention in the game Trivial Pursuit. Then a casual exposure to 19th-century scientist Michael Faraday's experiments got him pondering weightier matters. Specifically, he began to believe that on a subatomic level the universe was composed of gyroscopic particles. ("Just like little gears.") By 1968, he had developed the theory to his satisfaction, without the cumbersome baggage of mathematics or a formal education.

And there his troubles began.

"In about 1978," he said in a monologue that cut rapidly from chase to chase, "the devil came to me. I'm a very good scientist. I document everything I do, and I still got a copy of that. It was about five o'clock in the morning, and I was dreaming. This entity flew in front of me and offered me anything that I wanted if I would simply worship it. I immediately hit it in the face. It bounced like a rubber ball and came right back up, and when it came up it was always smaller, like a boxer hitting a punching bag. It dawned on me that you can't hurt this thing. You've got to outsmart it."

Upon waking, Newman found himself immersed in a palpable fog of evil. He searched the house, found the nefarious entity in the laundry room and killed it. That accomplished, he sat down and hypothesized that either an intelligent life form somewhere in the universe had beamed an electromagnetic force to him or that, yes, there was a strong possibility that God existed.

Newman then started praying for a sign regarding his energy technology. During this time, he was studying the planets and making calculations on huge sheaves of drafting paper. One day, while hard at his task, the number 14 filled his entire view. Not like an obsessive thought. It was literally the only thing he could see. "And all the major events of my life for the next seven years occurred to that number," he said, citing numerological connections to the purchase of a new house and other important events. In a fashion I was not able to deduce from our conversation, the number 14 also became instrumental to his scientific progress.

The world began to hear about Joseph Newman, as much as it was listening for him, right about 1986, when the Patent Office rejected his application for a device that contained 116 nine-volt batteries, a large coil wire, a magnet and a commutator (which repeatedly turns the power supply on and off and reverses the flow of the battery power). This device purported to produce more energy than it used, but the Patent Office concluded that the repeated stopping and starting of the power had caused misleading spikes to show up in the readout. Newman countered that the examiners had not tested his machine properly.

The story might have ended there, beached on the shores of ill will and frustration, had Newman not taken it upon himself to sue the Patent Office. He lost the case—no surprise—and added the defeat to his highly developed cosmology.

"This is a conspiracy against the American people, not just against Joseph Newman," he told me. The courts of this land, to his mind, are in cahoots with the Patent Office, as is almost every other large institution. His shorthand for the lot of them is "the power brokers." As in, the power brokers of this country who stand to lose everything if his invention sees the light of day and replaces every motor in existence.

But if his legal aspirations were dashed, the best was still to come. Soon after the court was adjourned, Newman received word through a visitor from "up north" that God had chosen him to lead humanity. He scoffed at the idea until a year later, when the Lord spoke to him personally. "This was not in words," he cautioned. "If it had come to me in words, I would have thought the government was putting something in my house or was transmitting electromagnetic waves to me. God spoke to me mentally."

The message was, quite logically, that the end was near, so Newman rented a coliseum in Mobile, Alabama, to break the news to the American people. For his efforts, he received responses ranging from disinterest to disdain. It wasn't for lack of timing, though, since his energy technology is scheduled to go into production on August 22 and the apocalypse will commence on August 21, as a result of the recent Jupiter-comet catastrophe. And as everyone knows, a perpetual motion machine can come in mighty handy when the normal laws of electromagnetics have been suspended.

Newman is taking the apocalypse very seriously. Maybe this is why he has decided to forgo technicalities like patents. Why bother with such trifles when cataclysm is at hand, ready to bring down the whole of Washington, DC, and your machine works perfectly well anyway? At the moment, he claims to have a unit "producing horsepower for a battery the size of your thumbnail." He also has $2000 worth of food in his home.

Talking to Newman gave me the feeling of being trapped inside a Nick Cave novel at daytime rates. Of course, I've been trapped in a Nick Cave novel before and enjoyed the experience, but there was still one piece of hard information I wanted to substantiate. I had heard that Ray-O-Vac picked up Newman's technology because it charged batteries for exceptionally long periods of time. When I asked Newman about this bit of dope, he both confirmed and updated it. Right after they made a deal, he said, Ray-O-Vac received more government orders than it had received in recent memory. Newman, fearing the hand of the power brokers at work, asked for his technology to be returned to him posthaste.

"Now what do you conclude from that?" he asked rhetorically.

Who knows? Maybe it was a conspiracy. Maybe it was a weird fluke. Or maybe it just takes a helluva lot of batteries to keep those monkey shows running. And running. And running.

August 24 – 30, 1994

The Greenhouse Helmet

"YOU KNOW THE BIOSPHERE?" asks Waldemar Anguita, sitting across from me in a crowded cafe. "Well, my invention is like a mini-Biosphere."

The prescient gleam in his eye, the blazer with turned up collar, the graying hair tufted up in two devil points—these details might telegraph a willful quirkiness in someone else. But Anguita is not exactly your standard-issue eccentric. His study of martial arts keeps him fleet of foot, and more fully in charge of gravity than any average shlub. In fact, his presence matches his name to a tee: the warm opus gives way to a lyrical roll, then on through a flurry of influences, to land like a cat on a balance. You want good things to happen to a man named Waldemar Anguita. You want his plan to work.

Here it is, on paper, between us as we speak. No moving parts, no dense hedgerow of formulae. Just a transparent dome to be worn on your head, with provisions for potted plants to be secured on the inside. To the fore, an air filter. To the aft, the same. In short, the Greenhouse Helmet (Patent #4,605,000).

The choice of the word "greenhouse" might be a little misleading here. The idea is not to breed beauteous flora or health-inducing snacks by virtue of your handsome face. Nor is it to provide interesting scenery while you deepen your forehead tan. As the wearer exhales carbon dioxide, the plants exhale purified oxygen in return. Thus, the Greenhouse Helmet constitutes an improved air purifier.

"I don't remember exactly what I was doing when it came to me," he says, gesticulating in a manner that aspires to music, "except that maybe it had something to do with the screenplay I was writing."

Indeed, in his patent and his script one can see the left hand moving toward the right. The movie follows a Puerto Rican man who converts boxing footage to moving holograms as a novel training technique. Through circumstance, he becomes a boxer in his own right and goes on to win the big, good fight. But *Rocky* this movie is not. During his jogging regimen, the hero doesn't just grunt and gleam and exhibit ungodly quadriceps. He also wears one of these helmets.

"Similar things are available," Anguita admits, "but they use an oxygen tank on your hip and a—" he mimes an intake tube. "I don't know about that. Nature did not mean for us to breathe only oxygen. We don't know, for example, what kind of tiny particles plants give us. Scientists can't even find what the smallest thing is. First it was the molecule, then the atom, subatomics, and now I think they have found something even smaller. It's like—you remember that movie from the '50s, *The Incredible Shrinking Man*? Well, he's still shrinking."

No prototype for the Helmet exists yet, partly because Anguita would first like to wean his screenplay into the world at large. When a model does get built, however, there are certain things he would like to see happen. The intake filter, for one, should be made of metal screens stacked in layers, because metal, he says, cools the air passing through it. A cactus would also provide cooler air, according to Anguita. For a jogger ensconced in a sun-baked terrarium, this can only be counted as good news.

Normally, I would say that our conversation digressed as the evening wore on, but with Anguita, there are no digressions. Or rather, there is only digression, and all arcs cleave to their own particular logic. In a giddily widening curve, we progress from the subject of sidewalks ("We have to get rid of them, because the earth is a magnet") to Madonna's running shoes, which he recently bought at an auction for $4500 and on to the pressing matter of houseflies.

"I have developed a technique with my martial arts knowledge," he tells me, his hands feinting and darting about the tabletop. "They come one time, you move a certain way. Then they come again?" He halts an imaginary insect dead in its tracks. "I showed my friends. It really works. I can make all the flies leave my apartment."

You may have decided by now that this Anguita fellow is in need of a few extra marbles. Having taken to him so quickly, I was trying to salvage a few shreds of objectivity myself when, as if on cue, an outside opinion came to the rescue: we spotted two bicyclists parking outside and removing, of all things, their dust masks.

Bingo. Anguita waits for them to come in, then proffers a drawing of his invention. A quick question only. Would they use it? Baby-faced and brimming with youthful intelligence, the guy scans the page. "Definitely!" he declares at last. "Oh yeah. It's so obvious!"

Then the doubts kick in. Wouldn't the helmet bounce up and down and generally jiggle all over the place? And how fast exactly do plants process oxygen? Anguita offers him assurances that these things can be worked out. The evidence before us is meant as a first step.

Our judge does another double take, flexes his face impressively, then, yes, he gives the Greenhouse Helmet two happy thumbs up. In the flush of newfound consensus, we speculate jointly about a human-sized, ball-shaped version for weatherproof jogging.

Anguita leans back, immensely pleased with the validation. He drapes an elbow over the back of his chair.

"Well, then. I think we can say it like this," he says, orating now. "Things are only going to get worse. They are not going to get any better. So people are going to have to use my idea"—he flicks the paper with an index finger and arches the eyebrow—"or *suffer the consequences.*"

Personally, I'm sold. But first, I'd like to learn a few martial arts moves, so I can rid my place of all those Incredibly Shrunken Men.

November 10–16, 1993

Where Ice-Nine Ended Up

ABOUT 30 YEARS AGO, a man was relating an idea to his younger brother. The man embellished his story, improvised a little maybe, until he found himself describing a fictional substance that could freeze water in an infinite chain reaction. Some years later, the brother, whose name was Kurt Vonnegut, introduced the idea in a novel called *Cat's Cradle.*

You may recall the gist of it. The military develops ice-nine for the purposes of lifting tanks from the mud but succeeds only in locking the surface of the planet into a single block of crystallized muck. "It turned out to be a terrible idea," said Bernard Vonnegut when I called him at his home in the Albany, New York, area, "because there was no sex in it."

Sex appeal or no, Bernard Vonnegut ended up making his life out of water and ice. And air and rain and tornadoes and all that. In short, he became a meteorologist, a career that eventually landed him a professor emeritus title at SUNY-Albany. Not content to keep his nose buried in the academic feedbag, he also found himself exploring the frontiers of meteorology now and then. He invented a new kind of anemometer (which you and I know as a wind speed monitor), for example, and he has nursed an abiding interest in lightning.

Well, lightning is a good deal sexier than tanks in the mud. And somewhere around the time that ice-nine got its name, Vonnegut began to entertain an idea that is still pretty sexy today—the ability not just to observe lightning, but to cause it.

"Investigations of atmospheric electricity would be considerably aided," says the paper that Vonnegut submitted to the *Journal of Geophysical Research* in 1961, "if one could cause lightning to strike from a storm at will." The paper describes how he and his colleagues flew some balloons tethered to piano wires up into some thunderstorms (one assumes there was insurance) and discovered that they had to send the things up really fast; otherwise, the field around them became too weak. From this, they also surmised that lightning rods don't work very well, except in the extreme cases where the electrical field is just too strong to be mitigated.

So on they went and performed some indoor experiments with a Van de Graaff machine, a huge industrial mother from the age of hugeness. Inside this machine, they sent a wire into an electrical field at 20 meters per second. It triggered sparks about three meters long—pretty exciting stuff for a weatherman.

Soon after the contained bravado of these experiments, however, the lab grew quiet.

"About forty years ago," Vonnegut explained, "I came up with an idea that's different from what has been believed for almost a hundred years." Scientists generally agree that lightning is carried by electrically charged precipitation, he said. Vonnegut, on the other hand, thinks lightning is transported by electrically charged convection currents—basically, that lightning is carried to the earth by a down-pressing wind.

Vonnegut has also noticed some squirrely aspects to tornadoes. Before a funnel appears, lightning and thunder will do a contact improv all over the dusty earth, but as soon as the twister arrives, the lightning and thunder disappear. From this and other details of the trade, such as how clouds are charged, Vonnegut wonders whether tornadoes are actually a form of lightning.

Unfortunately, to entertain such notions is to incur the wrath—or worse, the indifference—of the Almighty Grantgivers, and Vonnegut has drifted more and more to the periphery of the prevailing scientific community over the years. He still wants to try his experiment up in the air where the lightning strikes *au naturel.* He's even got his second wind for fundraising lately. Not much luck so far, though. You know how it is: everybody talks about the weather, but nobody wants to pay for it.

SHOULD VONNEGUT ever bring the bolts to earth, Percy Jones could be among the first to know. Better known for his bass playing in the erstwhile Brand X, Jones has been paying attention to thunderstorms ever since he made personal acquaintance with one back in his native Wales.

"One day, when I was nine years old," he told me by phone, "I was reading a comic book on the farm where I grew up. There was a storm, and lightning struck our phone lines. My mother got thrown across the room. All the light bulbs exploded, the wallpaper peeled, that kind of thing. So that sort of got me interested in the physics of it."

From this interest grew his thunderstorm detector, which, I might add, somehow seems in keeping with a bass player's sensibility—trafficking in the low rumble and all. I got a chance to see the device for myself when I visited Jones in his East Harlem apartment. It had the kind of design you don't see much anymore, the look of something that somebody has actually invented. Inside a box covered with masking tape, and hand-drawn labels, he had placed a light-emitting digital dial that can register the distance of any thunderstorm currently doing business within a 100-mile radius. Alas, the day I was visiting, it was blue and sunny, and the detector displayed only its default setting.

"When I was working on this a few years back," he said, "I used to bring it to rehearsal studios. People would get a real kick out of it when it started to go off."

Jones told me he intended to add a direction-finding function to the detector when he had the chance. Meantime, his apartment was crammed to the gills with tinkerings of ambiguous nature: voltmeters, aerials, ham radio equipment. In a way, being in his living room was like traveling back to a recently perished era. I'd wager that only a small percentage of his possessions had a silicon wafer in their guts.

Indeed, the project occupying most of Jones' attention these days (aside from his music) is his AM radio antenna, hardly what you'd call cutting edge. But the man has his reasons. State-of-the-art AM antennas are incredibly large, putting space at a premium for anyone who wants to set up a radio station. Jones, however, has managed to design an antenna that is very small. It's not good enough for AM play yet, because the bandwidth is too narrow, but for the nonce it works fine for ham radio.

For proprietary reasons, I can't describe this marvelous piece of machinery, but I can say that Jones made sparks jump from it, and that he made it light up a freestanding neon tube—all the old-style stuff. I mentioned that his projects reminded me of the days when inventing was closer to stage magic. He agreed and said he had originally planned to sell the thunderstorm detector as a kit. "But no one seems to be interested in that sort of thing these days. You never see things like Heath Kits anymore."

He hung the antenna out his window—an outrageously easy act as AM antennas go—and wheedled his ham radio tuner until he started picking up voices. As we listened, I began to have a pleasant sensation. We were cruising the back roads of technology, miles from the hectoring hard-sell of the information highway. The dialects of a thousand cabbies rose up in the pulsing, blue Manhattan air. And there wasn't a storm in sight.

February 23 – March 1, 1994

Convert-Me-Not Systems

RAYMOND JOHN HOWARD used to walk up dark stairways, through dank cellars, down dubious alleys. Safety was never an option. Floorboards could fall, rats could come running amok. One false move and his head would hit the floor with a crack.

Such is the life you might expect to lead if you were a private eye, but Howard wasn't snooping for dirt. He was just another guy reading meters for Brooklyn Union Gas. "Eventually, I had to get out," he says, trolling his vegetarian chili for substance. "I mean, if natural gas were proposed today and it had never existed before, it would never be approved. It's just so dangerous."

Brooklyn Union Gas was an odd place for someone like Howard to be spending his time in the first place. He's a thinker by nature, with a far-reaching scope and a collegiate sense of delivery. An architectural draftsman, he took the meter-reading job because work at the drawing board had hit a slow patch. But apparently, even the basements of Brooklyn have their lessons to teach, because they got him thinking about a set of inventions that could eliminate gas altogether.

"Some scientists," he says, by way of introducing his idea, "have become concerned about the large amounts of methane in the atmosphere, because it contributes to the greenhouse effect. Well, I read a figure that one seventh of all natural gas gets lost to the atmosphere. And that's not during distribution or in people's homes. That's just in leaks. If you add it all up, I think you'll find a good part of the methane-warming problem right there."

Having returned to the drafting beat with a conviction that there had to be a better way, Howard began to formulate a thought experiment. What would happen, he wondered, if the existing forms of energy kept the same form from source to consumption? Could the mechanical energy of water and wind be used in mechanical products? What if heat never turned into anything else, and light was light from start to finish?

The notion seems curiously appropriate for a man with three first names—as if the conversion from "Howard" to "Howardson" involved a loss of its own. Then again, as long as Sony can trot out its perfunctory "Sony Wonder" exhibit (in which interactivity means nothing more than a spectacular elaboration on the phone-machine maze), it's also an idea worth considering.

Certainly, the world as we know it today has few non-converting systems to show for itself. Gasoline explodes for the sake of pistons and camshafts. A hydroelectric plant, as the name suggests, converts "hydro" to "electric" all the livelong day. And nuclear energy would be no big shakes if it simply stayed in the nucleus.

Adhering to the no-conversion rule, then, Howard began with various energy sources and followed them to their logical conclusions. Windmills, he figured, could be used to liquefy air, which could then be pumped into cars for fuel. Waterfalls could drive pneumatic household products through the gas pipes that are already in place.

Because the sun emits both heat and light, it's especially powerful in Howard's scheme. He imagines a worldwide network of solar banks, with half of them taking in the rays at any given moment. But his massive solar-energy web wouldn't serve as an electrical system. That would be cheating. Instead, it would have twin streams of fiber optics extending to every point on earth—one to relay heat, the other to send light. For a bonus, the same points that generated energy during the day could turn around and receive it at night.

Tantalizing as these proposals were, one thought kept nagging at Howard's mind. Electricity, he felt, was ideal for many processes, with electroplating, welding and wireless technology ranking high on the list. But how could he get electricity in its natural state? Well, as it turns out, the answer came from "The Patent Files" themselves. He simply read about Dr. Bernard Vonnegut's attempts to provoke lightning by sending rockets into the air, and the idea came to him—you know—in a flash.

It's a description befitting an ark, or at least a covenant. Build a structure about the size of a football field, says Howard, by stacking thousands of sheets of glass. Enclose the entire monster in glass to prevent the current from leaking out. Then run a rod from the center of the building into the earth for a ground.

Such a structure would resemble a giant Leyden jar—a device that can hold an electrical charge. Benjamin Franklin, for his part, used an ordinary-sized Leyden jar when he flew his intrepid kite, and in some ways, Howard's plan is Franklin's experiment writ large. Departing from our frisky forefather's method, however, Howard wants to bring the charge down with balloons. Big balloons, with metal skins.

Picture a kind of ballet. One balloon would act as a kind of messenger, roving the sky when all was clear. When a storm approached, this messenger balloon would register a charge in the atmosphere, causing a fleet of "work" balloons to rise. These work balloons would then act as lightning rods, attracting the charge into the glass-and-foil gargantuan. After the storm, the process would simply reverse: the work balloons would return to earth, and the messenger balloon would continue its sentinel task.

Putting this scheme into practice presents a few difficulties, of course. The balloons would need some pretty thick hides, so as not to suffer the blows of outrageous fortune, and some pretty fast semiconductors would have to be installed to keep the voltage from draining away. With a million volts at a time coming in, steps would have to be taken to break this dose down into smaller units. And from a builder's standpoint, the biggest challenge might well be finding a way to install those sheets of glass without breaking them.

"Remember," Howard cautions, our meal behind us now, "this is just a thought experiment. Maybe it's not really practical." Pause, clatter of dishes offstage. "But I would like the opportunity to find out."

Well, the key to finding out, as always, is to make the proposition attractive to those with means. I think I've got just the ticket here. Assuming (as Howard does) that the flagship could be built in central Florida, where the storms just keep coming and coming, why not ask Disney to foot the bill?

Imagine looking out of your window during a thunderstorm. Off in the distance, a whole fleet of balloons ascend, a veritable Thanksgiving Day Parade of them, with the faces of Grumpy and Sleepy and all the rest painted onto their skins. As the lightning bolts converged on the wires in a majestic conflagration, these cartoon effigies would smile away, their visages undiminished by the powers of nature. Then the sky would clear, the balloons would slowly recede, and the sun would shine down on that lone tracking balloon, which would bear—what else?—the face of Snow White.

Okay, so maybe it's not art. But it's got to be better than forcing people to trip through the basements of Brooklyn.

January 25–31, 1995

The Riddle of the 100-Mile Carburetors

SOMETIMES, WHEN I tell people what I do, I mention the 100-mile carburetor, just to test the waters. Occasionally I meet a skeptic, like the shelter designer in West Virginia who dismissed this invention as a trumped-up rumor. For the most part, though, people do everything short of crossing themselves. An insanely good-looking New Yorker asked me if I thought $374 trillion, a reference to the annual revenues from gasoline in 1992, was a motive for murder, then asked me not to use his name. A solar-power engineer from Maine was uncertain: "You mean the patent they've got locked up?" A sanguine Ohioan applauded my interest, then added, "If you write about that, you could find yourself wearing cement overshoes."

The legend of the 100-mile carburetor is pure Americana: it combines wild hopes, technical know-how and paranoia into a single unusable mass. It's both everywhere and nowhere, a spooky tale of Yankee ingenuity crushed by corporate evil. And like all resilient conspiracy theories, the maddening lack of proof is built into the myth.

The story behind the legend, if the two can be separated, traces back to a Canadian inventor named Charles Nelson Pogue. In the 1930s, Pogue took out a series of U.S. patents for vaporizing automobile fuel before it entered the engine cylinders. "I have found," he wrote in Patent #1,938,497, "that in [ordinary] carburetors a relatively large amount of the atomized liquid fuel is not vaporized and enters the engine cylinder in the form of microscopic droplets . . . The remaining portion of the liquid fuel which is drawn into the engine cylinders and remains in the form of small droplets does not explode and thereby impart power to the engine, but burns with a flame and raises the temperature of the engine above that at which the engine operates most efficiently, i.e., from 160 to 180 degrees F."

If these droplets of fuel could be vaporized, Pogue reasoned, cars would get much better mileage. He tried to achieve this vaporization by heating the fuel at the carburetor stage; hence the title of convenience, "the 100-mile carburetor."

Pogue took out at least four patents, and they can be seen at any Patent Depository in the United States. But the mythology attaches another patent to his life work. This rogue patent allegedly achieved 200 mpg, and some assume it was appropriated by the Big Three or the government—or both. This is "the one they've got locked up." As for Pogue, little is known of his life, which only bolsters the conviction that he somehow got locked up, too.

Still, Pogue's efforts were not in vain. Like the human mind, which has some claims to existence because it occasionally appears in the human brain, the dream of outstanding mileage continues to make its cameos in official documents. In 1958, Robert S. Shelton received Patent #2,982,528 for a "Vapor Fuel System." In 1979, Thomas H.W.W. Ogle patented a "Fuel Economy System for an Internal Combustion Engine" (#4,177,779). Other vapor-fuel systems go through the grapevine from time to time. Recently, a patent lawyer told me that "two guys in Cleveland" had made good progress on one.

That the 100-mile carburetor has left a paper trail at all gives it a unique status in the world of invention. No patent examiner, for example, will allow the words "cold fusion" to appear in an issued patent, because the notion falls beyond the pale of known physical laws. As a result, cold-fusion hopefuls learn to camouflage their claims, hoping that they will prove useful in the event of a breakthrough.

Vapor-fuel systems are not like this; they have appeared without any disguise whatsoever. Shelton asserted in his patent that "approximately eight times the mileage that is obtained by the conventional internal combustion engine is provided" by the use of his system. Ogle's patent was more specific: "Due to the extremely lean fuel mixtures used by the present invention, gas mileage in excess of one hundred miles per gallon may be achieved." True, these numbers probably represent peak performance under clinical conditions, but a car that gets 100 miles to the gallon some of the time is still a fairly amazing thing.

Whether the car companies are holding on to any of these vapor-fuel patents is another question, and in the absence of proof, one can only gaze into the bureaucratic entrails for clues. Certainly, the needle is hitting empty later than it used to. My mother, a cautious source if there ever was one, says her Honda Civic recently got about 43 miles to the gallon on the open highway. Volkswagen already has two 56-mpg cars out in Europe, and has announced the introduction of a 90-mpg wonder by the turn of the millennium. As with computer chips, we may be getting our harvest feast morsel by morsel.

Then again, you could look at the thing from another angle altogether. Conspiracy theorists invariably think of the 100-mile carburetor in totalitarian terms. The Big Three, the story goes, are working with the oil companies to suppress this invention because inefficient cars mean bigger profits. This theory frames the debate in terms of mass production. But maybe the idea of mass production is obsolete in this case. Maybe, when it comes to vapor fuel, it makes more sense for people to build 100-mile carburetors themselves, as a one-time-only custom job. After all, $374 trillion is only a motive for murder if you have that much on you.

If *The Secrets of Super Mileage Carburetors: How They Work How to Build Them* is any indication, Wayne Heldstab would agree. You won't find rabid proclamations of miracles in his homemade primer, nor will you get disparaging remarks about lunatics. Instead, Heldstab gets straight to the point and tells you how to do it yourself. And while he lays out the basics of 13 inventions—some of which have been patented and some of which have not—he by no means stretches any of the inventors' claims. "No guarantees are made that better mileage can be achieved with these devices," he writes, "but it is a known fact that the internal combustion engine was originally designed to burn with fuel vapors."

Ironically, it is also a known fact that vapor-fuel systems are suited for the worst cars. This is partly because the fuel-injection systems and computer-assisted carburetors in newer, more efficient cars tend to snake throughout the chassis, while the carburetors in the old gas guzzlers tend to be almost modular. Thus we learn that the "Fuel-Mizer" allegedly got 100 mpg in a Lincoln, and that Ogle's carburetor achieved the same figure in, of all things, a 1970 Ford Galaxy.

Admittedly, developing a vapor-fuel system is trickier than growing live sea monkeys, and Heldstab is right to identify his reader carefully. "If you can fix most anything on your car without an instruction book," he explains, "and you can build toy furniture and gadgets from scratch without pictures and instructions, you probably do have the imagination required to build your own carburetion system."

I would second that opinion, with one reservation. You *should* try this trick at home, assuming you fit the above description, but I would strongly discourage any attempt at commercial production. It's awfully hard to run a factory when you're wearing cement overshoes.

August 30 – September 5, 1995

Out, Out, Damn Radiation

IT MAY SEEM like writing this column is nothing but a romp through whimsical mindscapes, but sometimes a situation will come along to make me feel the touch of darkness. After I described the circumstances surrounding Yull Brown's invention to a friend, he told me that the whole shebang had been poisoned somehow and shouldn't be published. Another suggested that Brown was trying to horn in on the cold fusion race. For my own part, I feel as if I sent away for the X-ray specs in the back of a comic book and ended up in a shadowy maze.

My journey began with the July-August-September copy of *Extraordinary Science*, a publication put out by the International Tesla Society that explores the fringes of scientific fact and runs a disclaimer admitting as much. Leafing through it, I came across an article by Yull Brown, who holds two U.S. patents for what he calls Brown's Gas (#4,014,777 and #4,081,656).

The pictures here are certainly calculated to raise an eyebrow. One of them shows Brown, bald-pated and frosty-bearded, welding glass to brick—not exactly sexy, but a neat trick nonetheless. Another reveals Brown behind a contraption described in the caption as a "high volume generator" that purports to produce 10,000 liters of gas per hour.

In the text itself, Brown starts by describing his method, which separates water into specific proportions of hydrogen and oxygen, producing a flame upon recombination. He's welding, using water as fuel. While unusual, the process does not depart from known science, according to a physicist who looked at the patents at my request. Rocket fuel, for example, works according to a similar principle.

But soon enough, things get a little hairy. Brown, not content to leave his invention within the bounds of proven fact, proceeds to announce that in the last few years he has used his gas to—and no snickering here—reduce radioactivity.

Now, I'll admit to letting myself be drawn in by this. Heaven forbid I should get any definitive answers about Brown's Gas, seeing as my knowledge of nuclear physics is about as broad as the few working nuclei I have left. Instead, I became absorbed in the dramaturgy of the persuasion—how it worked and how it failed. I suspended my disbelief, you might say, for the sake of an education.

At the heart of Brown's article is a statement written by Daniel Haley, a former State Assemblyman from New York, concerning a demonstration conducted for the Department of Energy on August 6, 1992. During this demonstration, Brown is said to have reduced the radioactivity of Cobalt-60 from 100 to 4 percent. The D of E later admitted to Haley that the readings said as much, but concluded that the radioactivity had either gone farther inside the material (in a process known as encapsulation) or else had "changed shape."

At this point, a man named Bob Dzajkich, of Southwest Concrete Products in Ontario, California, enters the story. Dzajkich was ostensibly interested in Brown's Gas for the purposes of welding, but upon hearing of the D of E's conclusion, decided to investigate this radiation business further. He conducted the same experiment, then crumbled the material to dust. When the Geiger counter still gave a reading of 4 percent, Dzajkich became convinced that the radioactivity had in fact been reduced. And so, from the looks of the statement, did Haley.

So far, the emotional slant is that the D of E is myopic—certainly no novel observation in itself. Only one problem: nowhere in his statement did Haley claim to have been an eyewitness to the proceedings.

My phone was just sitting there, looking like an expectant pet.

Dan Haley is about as pleasant a man as you are ever likely to meet. When I called him up, he said he thought the people from the D of E were "nice guys, probably in their early thirties. If they say something, they're going to have a strange mark on their careers." As for himself, Haley said he began his career as a champion of alternative energy, then gradually became involved in the oil industry and now lives in Dallas. (Note that he isn't a scientist, so he wouldn't be able to verify the results of a radiation experiment, witness or no.)

"We live in an age of great invention—and you would never know it," he told me with boyish charm. He also gave me Brown's phone number, with the tip that a scientist had recently asked him to duplicate the radiation experiment, only to have Brown turn him down.

"Yull can be ornery," Haley cautioned.

As it happened, Brown was not at all ornery when I called him at his home in Encino, California. In a soft voice that processed his native Bulgarian syntax through an Australian accent, he gave me the dope on his alleged feat.

"If you take a welding torch—oxyacetylene—and heat it," he said, "it takes dark glasses. If not, you get a complication. That is the reason that everywhere you go welders wear dark glasses. With my flame, you don't need glasses. Why? Because my flame is implosive, not explosive. And the infrared and ultraviolet rays which are produced when you heat material can't reach your eyes, because the implosion pulls it inside on [sic] the melting point."

"It's like a black hole," I suggested. "It takes the light in."

"Exactly!" he replied. "Now, all nuclear scientists will say that if I take a radioactive material and heat it, the radioactivity will go into the air . . . But with my flame it's different. Instead of the radiation going out, it's pulled inside on the melting point . . . Because of extremely high temperature, it goes directly inside the material and loses the radioactivity."

I wouldn't say Brown presented himself as a stunningly happy man, all told. He mentioned several times that the Trilateral Commission was trying to stop him, and on one occasion suggested he might meet "a not so normal end."

Most likely, this worldview has not helped his cause very much. When I asked why he had turned a fellow scientist away, he said, "I want to make a good business. I want to give it to the people. So why should he start where I am?" When I asked how the coming generations could ever hope to recoup his knowledge, he simply told me how ardently he believed in the coming generations.

Let's take stock of this situation. Certainly, anyone possessing the ability to reduce radioactivity would stand to become immensely rich. Then, too, there is that niggling matter of saving the world. So the question boils down to this: why, in the absence of scientific validation, would a former State Assemblyman—who is now affiliated with the oil industry—support Brown's claims about the reduction of radioactivity, mingled as they are with dimestore conspiracy theories?

Good question, Dave. Among the facts at my disposal, only one seems to stand out, not as anything like an explanation, but as a dim beacon in what has amounted to a surpassingly dense fog: Brown's first U.S. patent runs out on March 29 of this year—mere weeks from now. After that, the technology is likely to enter the public domain. So who knows? This story may well belong in the crackpot bin, but if you hear of any further developments about welding with water with a different name attached, don't say I never told you.

February 2–8, 1994

A Sign of Another Time

WHEN 40TH STREET and Sixth Avenue became Nikola Tesla Corner, the earth did not split in half. Pedestrians plowed through the workaday midtown melee, too hell-bent on the future to take much heed of the past.

I'm not sure what I had expected. The year marking the centennial of Tesla's most astonishing output, I suppose a salvo of electric fireballs over Times Square or a Medusa head of lightning on the Empire State Building would have struck the appropriate note. But no. There was only this small sign, looking like a municipal afterthought.

The devout will be quick to take the slight to heart. Before Tesla's career was even cold, in fact, the official version of the Serbian-genius-as-underdog was already taking shape. The inventor helped create his own mystique, of course. He seemed to enjoy playing the martyr—with his 18 napkins at every hotel meal, his vacuum of a love life, his tales of a numinous childhood.

And yet, when he wasn't preening his image as a fin de siecle mad inventor, Tesla was busy engineering large sections of the American century. His most lasting contribution, for one, should have put his name on the top of your monthly electric bill. Not that I'm upset about it, but seeing how his most ardent followers are still trying to convince the world that he was from Venus, the time seems ripe for a few dispatches from Earth.

Tesla had his share of rivals in his day, but none was more formidable than Thomas Edison. When the two first met, Edison was heavily committed to direct current electricity, even though it required that the power source be very close. As a result, Tesla's proposal for a cheap and practical AC generator—the first ever—didn't go over too well, and Tesla had to find another backer. Enter George Westinghouse with resources in hand.

The battle that followed, known as the War of the Currents, would have had latter-day politicians blushing with envy. Seeking to put the competition in a bad light, Edison set about electrocuting stray pets with AC electricity and then rushed to the press to declare them "Westinghoused." One of Edison's assistants went so far as to secure the rights on Tesla's AC patents for the purposes of Westinghousing one William Kemmler, a death-row inmate. It took some time to do the job. As patent historian Allen Koenigsberg described it to me, "It wasn't really an execution. It was cooking."

The tides began to turn in 1893, however. The Columbian Exposition saw Chicago transformed into a blaze of AC-powered lights, under which Tesla demonstrated a multitude of his inventions to slack-jawed crowds. In October of that same year, the Niagara Falls Commission agreed to use Tesla's generators. AC had finally gone legit.

It was quite a year all around for Tesla, actually. Fluorescent lights, a prototype of the cyclotron, an electric synchronized clock, electric oscillators—everything seemed to pour from his mind at once. Contrary to the social consensus, it was also the year that he demonstrated one of the world's first radios. The Supreme Court handed down its ruling on *Tesla v. Marconi* in 1943, mere months after Tesla met his maker. Doubters can look it up.

But Tesla never really passed in the circles of high finance. His imagination tended toward pressing matters like affordable energy for everyone on the planet, and most people just wanted to get rich. The science of robotics and an early version of radar, both attributed to Tesla, fell on the stone-deaf ears of Navy admirals. His lifelong ambition to transmit power without wires proved too expensive for the likes of J. Pierpont Morgan. In the end, his reputation led him on a downward spiral of improbable ventures.

And it led him to Bryant Park, where the street sign bearing his name now stands. Having fallen out of step with the times and deeply into debt, Tesla developed a close attachment to pigeons. No one knows why exactly, though many think his mother had something to do with it. At any rate, it soon became a common sight—the inventor in the park, ministering to his flock.

One day around this time, Tesla was invited to receive—of all things—the Edison Medal. Incensed at first, he eventually bit the bullet and agreed to appear as the man of honor. The shindig proceeded apace. When the time came for him to accept his prize, however, he was nowhere in sight. Someone thought to look in Bryant Park, and there indeed they found him, covered from head to toe in a living carpet of pigeons. Dusting himself off, he returned to receive the award with his customary grace and charm—and his customary reluctance to explain his affairs.

You could say Tesla didn't have his marketing strategies totally worked out. The durability of his type is another story. As I looked up at the street sign, pondering the wages of brilliance, I noticed a man standing next to me. It seemed he was looking in the same general direction, so, ever the fool, I asked if he was a Tesla fan.

"My teeth," he replied, pulling back a lip to reveal a ruin of molars and pink mash, "are synthetic."

And with that little pearl, my witness wandered into the pulse-fed logarithms of Times Square, regaling an invisible choir.

The American century is certainly having its troubles. Now on the heels of everything comes the news from Paul Brodeur in *The Great Power-Line Cover-Up* that high-powered electromagnetic fields, of the sort found around the power stations Tesla brought into being, could be giving cancer to those who live nearby. And, if I may take the metaphorical leap, I suppose that means that you and I and the man with synthetic teeth have all been Westinghoused, writ large, while the lights of Broadway shimmer.

Only now, of course, the engines of injury all bear Edison's name.

October 27 – November 2, 1993

Sensible Alternatives

THESE DAYS, as the most trivial conversations become fodder for legislation and movie rights precede their real-life stories, the words of Mark Twain rumble roughly through my mind: "When you have the chance," he said, "do the right thing. It will please the right people and astonish everyone else." And so, at the risk of pleasing Mark Twain, I submit for your approval two inventors who are behaving in a truly shocking manner—they're trying to make the world a better place.

After working on some lesser inventions (a Post-it holder, for one), Winston MacKelvie of Knowlton, Quebec, became interested in potentials for recycling energy. With so many people collapsing into self-indulgence, and so many machines being so inefficiently built as to give new meaning to the term *waste product*, he certainly had his choice of fields. But every inventor must find his obsession. Perhaps inspired by his wintry climes, MacKelvie chose to tackle the heat that's lost when ordinary hot water drains through the plumbing.

Those who have meditated for any time on this subject have invariably run into a daunting obstacle. Capturing heat from draining water generally requires heat exchangers, which absorb the heat and let the water go on its merry way. But if cold water drains through the exchanger—when you splash your face in the morning, for example—it "steals" the stored heat, and nothing is furthered. When MacKelvie sat down and looked at the state of the art, the best he could find was a device that would channel any water that was 80 degrees or cooler away from the heat exchanger altogether. Pondering this, he thought he saw a better way.

Like many good ideas, DrainGain is stunningly simple at its core: it takes advantage of the fact that heat will always rise to the top of a body of water. By putting a heat exchanger (shaped like the coil on an electric stove) at the base of a specially designed water tank, MacKelvie has ensured that the heat will always be introduced in the colder-water zone. This heat then rises to the top of the tank, where another exchanger takes it out of the tank and routes it into the ordinary hot water supply.

MacKelvie has devised two other components for his system: a method of routing the solid wastes (pretty damn important), and a dimpled tube for his heat exchangers, which creates enough turbulence to keep the finer solids from caking up the works. Anyone who has ever seen a humidifier take on the aspect of an abandoned lime quarry will doubtless appreciate the practicality of keeping the tubes clean.

So what does DrainGain mean in the larger scheme of things? Well, most immediately, it means that water coming down the pike at 79, 74 or even 34 degrees can be put to good use, as opposed to the earlier system, which can only use water that's 80 degrees or higher. Zoom out a little and that could mean up to 50 percent savings on energy bills. Proceeding apace, it could make changes that are visible by satellite.

"If it really does save that much," says MacKelvie, who has a Canadian patent and a U.S. patent pending, "and it really can be built in a home for $500, and it really does keep Quebec from investing $13 billion for a new dam, and it really does keep that dam from flooding Native American lands, I don't see that there's another way to go."

Of course, even a superhero inventor has to look at the context. Copper, the material of choice for heat exchangers, would have to be mined more extensively in order to produce DrainGains for the masses. Depending on the number of units sold, the cost and the damage incurred when the copper is mined, there could be more drain than gain writ large. On the other hand, if we really do intend to become a nation of electric-car drivers by the turn of the millennium, we're going to need a hell of a lot of juice to keep the batteries charged, and any reduction in energy output elsewhere is going to help.

So many factors to weigh. If Rudolf Gunnerman has his druthers, for example, we won't even need electric cars. The latest developments at his company, A-55 Ltd., of Reno, Nevada, are under wraps for the moment, and will remain so until a secret business deal finds its level. Depending on how the proceedings go, Gunnerman will either make front-page news or live to see his brainchild wither on the Western plains. Meantime, dribs and drabs of the goings-on have reached the public record.

Like MacKelvie, Gunnerman is trying to minimize the amount of energy that has to come out of the ground. Unlike MacKelvie, he's working in the glitzier field of alternative cars. Either way, though, his A-55 fuel (Patent #5,156,114) has a lot of people scratching their heads as to how a car could run on 55 percent water, 45 percent gasoline and a smidgen of surfactant, otherwise known as soap.

Indeed, Gunnerman himself doesn't know how it works. The important thing is that it does—at least some of the time. Test runs have been conducted with varying degrees of success, and a hodgepodge of investors have been sniffing around with equal parts suspicion and interest. Of course, the obligatory Bizarre Science demonstration has been trotted out as well. Seems you can train a blowtorch onto Gunnerman's fuel and it won't catch fire.

A-55 fuel, if it makes the grade, has several factors working in its favor besides thwarting pyromaniacs. For starters, once the engine is coated with a nickel catalyst and a few minor adjustments are made, an ordinary car will run on the stuff. The whole makeover shouldn't cost more than $2000, tops. Then, too, there are the obvious benefits of burning less gasoline. The 70 percent water-vapor exhaust should pass California's emission rules easily, and the mileage promises to be a reasonable 27 to 33 miles per gallon. The real clincher, though, will be the cost: A-55, by using 50 percent less gasoline, could represent an equivalent percentage in consumer savings. *Oil Market Listener*, a trade publication dealing in a variety of energy concerns, has been advising the government, in a tone at once both stern and pleading, not to tax the living daylights out of a fuel that could conceivably be sold for as little as 60 cents a gallon.

Gunnerman is onto the kind of good thing that shouldn't be ignored. Gunnerman and MacKelvie put together are onto the kind of good thing we need pretty fast, pretty hard and pretty permanently. I mean, save 50 percent on energy bills and 50 percent on gasoline costs and we could practically re-institute a middle class in this country. And if that should ever come to pass, somebody better call Ripley's, because it really *would* be amazing.

June 22–28, 1994

4

DIGITAL VISIONS

Squatting in Cyberspace

NOW WE KNOW. The Internet is not a superhighway, a nightclub or an aquarium. It isn't a game of go or an art installation made of styrofoam and menstrual blood. The Internet is a megalopolis, with all the thoroughfares, districts and slums of an earthbound, carbon-based city. We know this because cyberspace has already experienced its first high-profile squatter.

At midnight on August 31, 1996, *New York* magazine closed down its CompuServe forum, leaving a cursory farewell. Only problem was, the propellerheads who maintain the service forgot to remove the structure of the forum itself. Gutted of content and content providers, it was still there, its message base and libraries intact. And that was the situation when Jim Leff stepped in to claim the forum as his own.

"People were really freaked," Leff explained when I phoned him at his Astoria, Queens, home. "You have to imagine: It was like opening *Newsweek* and instead of 'Periscope,' finding 'Leffscope.'"

Leff, it must be said, lives well within the domain of alternative possibilities. His enthusiasm for food, for example, borders on the messianic. As a food critic for *New York Press* a few years back, he was known to sing the praises of some of the most disenfranchised restaurants on the planet. Once when he was dining out, he interrupted a talkative friend with a raised palm and the sober announcement: "Not right now, man. This food is talking to me."

His personal life, meanwhile, has the quality of a heavily plotted low-budget film, complete with cameos by amorous belly dancers, scheming old ladies and hipster musicians. Some of these acquaintances can undoubtedly be traced to his career as a jazz trombonist, but others are less easily explained. On the eve of a trip to Lisbon a few years ago, I picked up the phone and heard him asking if I wanted the number of a friend of his who lived in a Doge mansion there, and who, by the way, happened to be a count.

Leff is also very much a man of the digital world. He's been ranging around cyberspace for three or four years now, and thanks to what he calls his "sheer refusal to hang up and go away," his presence there is a familiar one. As testimony to his computer chops, he was recently made a sysop (systems operator) for the Bacchus Forum, a congeries of beer, wine and food enthusiasts.

All of these combined resources went a long way in predisposing Leff for his rank. Still, he couldn't fully inhabit the abandoned *New York* forum without technical support. In fact, for the first 15 hours, his forum was little more than a Usenet group trapped in a CompuServe body. And so it would have remained but for a sysop who offered to liberate the news-flash box and the message sections in the name of Leffdom.

From there, it was a free fall all the way. In place of the usual news flash, Leff wrote: "So, you thought I was joking, huh? Sorry, but New York Magazine, the sponsor of this forum, pulled out at midnite on September 1. They are no longer in any way involved with or responsible for this forum. In fact, this forum officially doesn't exist. It's not here. You're imagining ALL of this. This is an EX-FORUM!!"

Promising to run a "tight ship," he then replaced the titles of 21 message sections with new ones reflecting his personal concerns, including such chestnuts as "JL Wants to Buy," "JL Wants to Sell," "My Hellish Love Life" and "Lunch With Bob." In one section, he pontificated on Trappist beer. In another, he bemoaned his love life. Exercising his newfound powers, he appointed leaders for sections on hockey and port wine. And for six hours, CompuServers around the world had the unique privilege of logging on to a runaway forum.

You could debate exactly how transgressive the Jim Leff Forum was. For one thing, he didn't break any rules, strictly speaking. "The reason it was okay," he explained, "is that technically the forum didn't even exist." There was also something a little too easy about the whole thing. The responses, enthusiastic as they were, smacked of the breeziness that has become *de rigueur* for on-line chatter. One smart-aleck claimed that the forum looked a lot like one he had recently lost. Another declared it better than *Cats* and promised to return again and again.

Indeed, in some ways, the stunt had all the party-hat trappings of vanity-book blowout. But in other ways—most notably the ways that involve money—it was a different beast altogether. While the exact numbers are kept under wraps, the cost of owning a CompuServe forum is reputed to run to five digits. The reason the figures are so high is simple: unlike most Web sites, forums tend to be heavily traveled. And in Leff's case, that meant a much larger audience than most individuals can ever expect to get.

"I was really sad when it closed down," said Leff, "and not just because it had joke value." After all, where else was the arepa lady—a Jackson Heights street-cart vendor who enjoyed a section all to herself—going to get such attention? "For a brief few moments," he mused, "she was a topic of discussion for people around the world."

If Leff was inconsolable, the mood didn't last. On September 9, visitors to the Handyman Forum saw a message pop up on their screens announcing, "THIS IS NOW THE GARLIC FORUM." Again an abandoned shell was peopled with boisterous squatters, and again casual browsers were taken aback. Perhaps the most satisfying moment took place when one user unknowingly wandered into the forum to discuss a plastics company in Michigan. Immediately, his thread was crashed by a band of garlic Grouchos, each employing wilder punctuation than the last.

Though Leff never gained leverage over the message sections, he did manage a kind of de facto control, with sections like "Grow Yer Own" and "Vampire Dissent" flourishing in the ashes of sections originally meant for heart-to-hearts on lug nuts. And this time, he was able to hold the fort for nigh on two days before a section titled "Burying the Bulb" announced the end of the hijinks.

Just who blew the whistle on either of these pranks is hard to say. In the case of the Garlic Forum, it may have been the man who came looking for plastics. Or maybe, as one on-liner suggested, CompuServe began monitoring Leff's account number for suspicious activities after the Jim Leff Forum embarrassed the hell out of the company. In any event, his adventures in cyberspace have hardly been curtailed. On the contrary, they seem only to have led to new horizons. He's been offered a spot in the miscellaneous section of the MacUsers Forum and a chance to spread the garlic gospel in, of all places, the Architects Forum.

Meantime, hopeful hackers angling for their own on-line forums won't be prying any secrets from him anytime soon.

"I will take to my deathbead," he vows, "the name of the person who handed over the magic wand." Leff has a lot of stories to tell, and for each one there's another secret waiting to be spilled. But somehow on this one, I believe him.

October 9–15, 1996

The Book of Books

ONE OF THE GREAT benefits of writing about inventions is that you can think up crazy ideas and wait for others to do the work for you. In the past few years, for example, I must have written a dozen reminders to myself to design an "electronic" piece of paper. Naturally, I had no intention of following through on these musings, because I knew perfectly well that someone like Daniel Munyan would eventually come along and invent the Everybook, which, as he puts it, is "the first electronic book that has the appearance of a printed book, with the ability to download, store and display hundreds of publications."

Munyan has certainly been playing the industrious ant to my dissipated grasshopper. For some years, he eked out a fairly routine existence as an employee of the Pennsylvania Credit Union League. Then on April 5, 1995, while circling Detroit, he had what he calls a "vision." Marooned in the boundless sky, he jotted down six pages of notes and wondered to himself if anyone had thought these thoughts before. (Daniel, my brother.) From there, everything fell into place in a very short time. Before he knew it, he was knee-deep in patent attorneys, a board of directors and a company capitalized at $2.43 million in private stock offerings. That's one heavy payload to grow from six pages of notes.

It's also among the most vigorously thought-out inventions I have seen. Open the lightweight Everybook Dedicated Reader along its vertical hinge and two high-resolution LCD screens turn on automatically, displaying the jacket covers of all the titles you own. There are no switches, keyboards or pointers. When you touch one of the book-jacket icons, its copyright page and table of contents appear on the facing screens. From there, it's a matter of dragging your finger to the chapter you want. Lift your finger and the first page of the chapter materializes. Touch the outside corners of the screens—the pages turn. Most important, all of this comes to you not in the type faces of your average Mac or PC, but as photographic reproductions of the original book.

When I told friends about the Everybook, most of them scoffed and gave it the gimmick verdict. I disagree. For starters, there are the obvious environmental and economic advantages. Where paper books pollute, an electronic book is tidy and clean. Electronic publishing is also a comparatively inexpensive process, and as such, promises to bring retail prices down. And since titles will be available over the phone from Munyan's Everybook Store, the Everybook system will free up valuable real estate currently occupied by mega-bookstores, not to mention give small, independent bookstores the cachet they deserve as sources for rare, paper-based books.

But none of these improvements would mean much if the Everybook failed to preserve what is already good about books—and this is exactly where Munyan kicks into high gear. Rather than twisting the book to the demands of the computer, he has adapted digital technology to a proven form.

Indeed, Munyan harbors no great love for the tomfoolery of most computer geeks.

"While I respect the work of the young computer geniuses of the seventies and eighties," he says, "I do not share many of their values . . . Most computer people think the book paradigm is obsolete. They want only non-typographic content, which they display across a medium that looks like a stage, movie screen or television. They force people to read, deal with the eye strain of the scanning bar and develop keyboard or pointer skills just to turn a page. They add video, sound and animation via multimedia, just to keep us from printing the written page and reading it in a more comfortable form—like a book. The computer is for work, and for entertainment; it is not for reading."

If Munyan holds the book format in high regard, he's also intent on preserving the basic outlines of the publishing industry.

"I believe that publishers are the most important link between authors and readers," he says, "and that the intent of the electronic gurus to turn books into electronic content without publishers is misguided at best."

To this end, he plans to encrypt each chip so that it can be read by one machine and one machine only. If anything, his electronic books will be harder to pirate than today's paper-based counterparts are.

Of course, as a writer, I'm duty-bound to applaud anything that furthers my own continued income. There is one area, however, where Munyan's distaste for the anarchy of the Internet gives me pause.

When I ask him about licensing plans for his patent-pending technology, he answers, in effect, that he has no plans to license at all.

"Publishers consign their publications to the Everybook Store," he explains, "much as they would consign books to a distributor or bookstore chain. The EBS serves as the central filesaver-based digital archive. Publishers need only send us exactly the same file as they would send to the print house, in the industry standard PDF file format. We can take it from there."

In other words, Munyan has a monopoly on his mind and, to the extent that the Everybook concept takes off, a distribution empire on his hands. If that's already a worrisome thought, Munyan adds to the mix by holding strong personal views: by his own admission, he's "an evangelical Christian" who reflects the values of his "conservative, agrarian community."

For the moment, those values remain firmly rooted in the political terrain known as the vital center.

"If you can get to a touchtone phone anywhere in the world," Munyan says, "I can get you a Bible, a dissident political publication or anything else that is not graphically pornographic or advocates criminal activity according to U.S. laws."

To illustrate the possible impact of his book-by-phone system, he gives an example that seems laudable enough. The Chinese, he points out, were unable to keep the Tiananmen Square massacre under wraps because the authorities needed the phone lines to talk to their generals.

Well, okay. But as always, the lesson of modern communications is to watch that wire, because for every case where a phone line acts as a conduit for freedom, there's another where it acts as a censoring device. Who will decide, for example, if a candidate for publication by Everybook is "graphically pornographic"? Moreover, will Munyan wax equally enthusiastic over a "dissident political publication" if the manifesto happens to dissent from U.S. laws? Somehow I don't think so.

The politics of the Everybook won't be a burning issue anytime soon. When the first titles are test marketed (at a starting price of $1600–$2000), they'll be aimed at a technical readership made up largely of doctors, lawyers and pharmacists. But Munyan is nothing if not far seeing. Eventually, he wants to break into college textbooks, and 10 years down the line, when the price of the Everybook is expected to slip below the $500 mark, his personal philosophy could figure highly in the ever-explosive syllabus debate. Will the maker of the book of books start pushing his own personal canon? Will he start playing God?

We know from recent experience that major distributors like Blockbuster Video and Wal-Mart feel perfectly entitled to alter the content of videos and CDs. Meantime, Munyan has a great invention and a great opportunity to give it to the world. As a professed advocate of a free press, he should put his money where his mouth is and license it to the highest bidders. Or I'll stop imagining things for him to invent.

December 4–10, 1996

Muybridge Squared

IN THE LATE 1870S, Eadweard Muybridge, a British prospector and photographer roaming the byways of the American frontier, happened onto one of the better deals of his era. Robber baron and racing enthusiast Leland Stanford had been talking with a friend when they got to wondering whether a galloping horse ever had all four legs in the air. To settle the question, Stanford hired Muybridge to take a series of high-speed photographs of a horse in transit.

Thus a career—and an industry—was born. Muybridge quit his wandering ways and, building on the work of Frenchman Etienne-Jules Marey, began to take sequential photographs of animals and humans in the midst of various activities. Some of these photos ended up in Muybridge's zoopraxiscope, a rudimentary projector that produced the fleeting illusion of motion. In 1888, Thomas Edison had occasion to see the zoopraxiscope for himself and promptly designed a peephole device called the Kinetoscope. A few years later, Edison's most capable assistant defected, taking with him the plans for America's first serious movie projector, and the cinematic age commenced in earnest.

I haul this story up from the vault because it seems to revisit the work of Dayton Taylor in more ways than one. For starters, Taylor's Timetrack camera actually resembles the zoopraxiscope on a technical level. Then there's what you might call future blindness. Timetrack may seem as insignificant an invention as the motion study did in its day; but with the right kind of serendipity, it, too, may prove to be an important step toward an entirely new technological genre.

In its present form, the effect is fairly basic. As with the zoopraxiscope, Timetrack uses a series of still cameras to record its subject. But where Muybridge tracked a moving subject through time, Taylor captures a motionless subject from different points of view.

Imagine, if you will, that a man sits in a chair, surrounded by a bank of 20 still cameras. Then imagine that the shutters of these cameras all open at the exact same moment, generating 20 photographic images. When these stills are then strung into a sequence and played on a video screen, they'll give the effect of a single camera "moving around" a photograph of the man. Run the film one way, and his face comes into view. Run it the other, and you return to the back of his head.

Not much to write home about maybe. But the possibilities suddenly become mind boggling when you do as Taylor's patent suggests and substitute video cameras for still ones. Now, instead of a series of freeze-frames, you have a series of moving images, each filmed from a slightly different perspective. The first images from each video camera will generate the same effect as Taylor's original setup: the freeze-frame pan. Consecutive images taken from one video camera, meanwhile, will deliver nothing more than your usual video footage. But use a combination of these two methods and a whole new realm opens up.

In fact, the full-fledged, video-based Timetrack camera promises not a single shot but a matrix of possible shots of the same event. Say our man stands up from his chair and blows a delicate, yet perfect, smoke ring. By linking the video cameras to a computer, this simple action could be viewed in any one of a number of ways. You could start with a view from the front, panning around to the man's back as he stands, then jump to the side as the smoke ring floats through the air. Alternatively, you could begin from the back and pan to a profile, or cut rapid-fire from side to side . . . in effect, you, the viewer, decide where the camera will go, even though the shot's in the can.

As if this weren't already enough, Taylor is entertaining even more permutations farther down the line. He originally designed his interlocking cameras to form tighter or looser arcs, and if he can get the film to behave when it bends, this feature may surface again someday. Elsewhere, his patent shows him stacking video cameras on top of each other, adding a vertical dimension to an already complex horizon. Then there's the distant dream of putting his contraption on a dolly, not to mention the cameraman's ability to zoom in or out—at any speed—on any one of these cameras.

Confusing? You bet. Yet for as much as Timetrack puts my head in a spin, what confuses me most is the personality of the creator himself. Meeting Taylor in the office of Electrokinetics, the R&D and engineering outfit that's taken on his project, I encountered the rare inventor who isn't obsessed to the point of distraction, who failed to bend my ear for hours, indeed, who didn't look terribly concerned whether I liked his invention or not. Well-groomed, well-spoken and nondescript of dress, he simply showed me a prototype, explained its workings precisely and fielded my questions without incident. If Timetrack flows from an overwrought brain, this man is hiding it awfully well.

I suppose it's possible that Taylor is too deeply absorbed by the vexing problems before him to waste his time with quirks. More likely, though, his nonchalance is a sign of his burgeoning success. In addition to the technical support provided by Electrokinetics, he's managed to rope in a handful of investors, including Steven Seagle, who writes for the *Sandman* comic books. He's landed a few advertising jobs from clients who like the effect. And thanks to the mediations of Roger Ebert, he's caught the eye of the potentate on high himself, Steven Spielberg.

Of course, what this wave of enthusiasm amounts to will depend a great deal on what Timetrack becomes. A letter from Spielberg to Ebert, included in Taylor's press kit, illustrates the point well;

between encomiums, the director finds himself wracking his brain "trying to think about applications for this art form/technology." Unless Taylor can suggest some meaningful reasons to use his brain-child sometime soon, it could easily go the way of technologies like Q-Sound. (And don't tell me you remember Q-Sound.)

Personally, I don't think it will, though, because it's the sort of invention that instantly suggests novel applications. I'll give you an example.

Four or five years ago, when I still suffered from the delusion that I was good with my hands, I had a vague plan to devise a musical recording that changed with every hearing. (If this sounds eerily familiar to you, stop here and write me a letter.) The idea involved no interactive button-pressing or joysticking. Instead, I imagined a randomizing component—a clock would do the trick—that would surreptitiously switch horns to violins, add an extra chorus, change the lyrics here and there, and so on. The listener would simply insert the CD and the music would come out altered each time.

Timetrack could achieve much the same effect with visual images. Every time you hit play, a video machine could choose its own camera angles, its own pans, its own cuts, creating different nuances and revealing unexpected details. Who knows? If a barometer did the randomizing, the pace of a Timetrack movie might even be made to match the weather outside.

Would people actually bother to watch a movie that made all the decisions for them? Maybe not, but you never know. After all, it was Edison—an accomplished Spielbergian in his own right—who described the Kinetoscope as "too sentimental" to interest the public much. And look what happened there.

September 10–16, 1997

Doppler 4000: What's in a Name?

NOT LONG AGO, I was watching TV with a friend visiting from overseas when he made the kind of observation that only an expatriate can. On first glance, he said, it looked as if *News Channel 4,* New York City's local NBC news program, was promoting its Doppler 4000 as a consumer product in its own right. *Weather machines!* the announcer seemed to be saying, *Get 'em while they last!*

What strange currents were these? Of course, I had long had my theories—involving, among other things, King Lear raging on the heath—but somehow my friend's remark struck an unusually deep chord. The histrionics of Penny Crone on a nondescript boulevard mere hours before the millennial storm suddenly seemed minor in comparison. Even the fact that *News Channel 4* itself promises the weather "as it happens" lost its sheen. From that night on, the Doppler 4000 took pride of place among my store of cultural riddles, edging out professional closet organizers and restaurants for dogs. In the end, there was nothing left to do but put in the call and find out what these people were selling.

The man who picked up the receiver at the other end of the line was Joe Witte, a meteorologist for 20 years and currently a morning forecaster for *News Channel 4.* See the Doppler 4000? Sure! Why not?

So up I went to the General Electric Building to see what I could see.

By all outward appearances, Witte was born to his job. He has the requisite smile, the captivating voice and, most important, the ability to give a plausible answer within three seconds. Certainly, he showed no aversion to laying bare the Doppler 4000 for me, despite being constantly called away to the set.

"Up until about five years ago," he explains, "we used radar based on the radio-tube technology of 1957. No one made radio tubes anymore, so it became very expensive to replace them. Then a new technology was developed."

Today, that new technology sits atop 30 Rockefeller Center, sheathed in a white fiberglass dome. In some ways, it's not so new at all. A 12-foot-wide dish that sweeps the horizon every 30 seconds, it uses the Doppler effect as most weather radar does: to determine the direction of raindrops in relation to the central antenna.

The difference, such as it is, lies in the power and reach afforded by solid-state technology. Stronger than its predecessor by a magnitude of 50,000, the Doppler 4000 can detect conditions as far as 300 miles away and is so sensitive that at times it confuses swarms of insects for precipitation. (Oddly, the new machinery still can't do much in the way of tracking snow, because of the relatively low mass of the flakes.)

The Doppler 4000 is not a must-have in the weather business. The National Weather Service—the government organization charged with keeping tabs on the rain, sleet and snow of the nation—has some 120 superior radars at its disposal. These radars, put out by Lockheed Martin, are scattered strategically throughout the country, and most local TV stations are content to rely on them for their weather, downloading data as they see fit.

Not *News Channel 4*, though. Looking for a competitive edge, it decided about five years ago to shell out for a solid-state radar of its own. About half a million dollars later, it had a dish from the Electronics Enterprises Corporation, a graphics system from Baron Services and a technical basis for its unusual boast of delivering the weather absolutely live.

"The National Weather Service has similar technology out in Brookhaven, Long Island," says Witte. "But it makes calculations for everything that's going on, and by the time it crunches all those numbers, the information is already about six minutes old."

The Doppler 4000, by contrast, can be marshaled to more immediate ends. It can isolate a specific area, scan it up and down and, say, zoom in on a discrete parcel of air—all while Witte is on camera. This kind of selective operation, called a *volume scan*, doesn't reach farther into the future, of course. It simply reaches deeper into the burbs, where the viewers are presumably strapped to their chairs with no clear view of the window.

"So what does 4000 stand for?" I ask, hoping to find some fact with some substance behind it.

"Oh, that. That's just a takeoff on Channel 4," Witte replies. "In fact, we had such a great promotion that other stations started to imitate us. Did you see our promotion last summer, during the Olympics?"

I couldn't claim to have seen this spot, but I had definitely heard about it. According to some people I knew, the thing lasted nigh on ten minutes.

"Actually, it was three and a half," says Witte. "But by the second time you saw it, it seemed like ten minutes. Anyway, what that has spawned across the country is a whole bunch of Doppler 7000s, Doppler 9000s, from stations promoting their software."

When pressed, Witte is only too happy to show me what this much imitated machine can do. Leaning over a desk, he puts electronic pencil to pad and the screen above him gives way to a closeup of the south shore of Long Island. A few more pen strokes and the image zooms in closer again, to the environs of a certain Sacred Heart High School, where some of the station's local data is collected. It looks like a clear day at Sacred Heart. I imagine Catholic schoolgirls in their woolen skirts, peering into the class thermometer, raising wet fingers to the wind.

Another non-starter. Then my eye seizes on one last promising sight: a small screen to the right, with the green spikes that look for all the world like an EKG readout.

"Oh that," Witte laughs. "That's just a fake dial reading the screen. It's just particularly for looks."

By now, I'm beginning to feel that I'm imposing on a man who is, after all, taking a fair amount of time out from his job to explain objects that don't really mean anything. Before I make my exit, though, Witte insists on copying the contents of a pamphlet put out by Baron Services. I'm glad he does, too, because more than anything else, this little primer gives me an inkling of what weather radar is really about.

Walking out of the GE Building into the glorious, sun-drenched day, I ponder the cone of silence, a "vertically oriented cone above the radar that is not covered by the main beam of the antenna." I cogitate on a "phenomenon by which the radar signal propagates along the boundary of two dissimilar air masses," called ducting. I linger especially long on a product known as the *storm total*, which describes the sum of the precipitation that hits the ground while the radar is trained in its path.

King Lear is nowhere in evidence on Forty-nineth Street. No raving fallen monarch bemoans America's "poor naked wretches . . . that bide the pelting of this pitiless storm"—not on a five-star day like today. Still, *storm total* isn't bad. If I were buying a radar system for a news show that likes to lead with the weather, I could easily go with a name like that.

May 7 – 13, 1997

Grudgers Redux

OBSERVING CURRENT EVENTS can be like watching a bloated river push against the levee. You look at it and wonder: how long before the flood?

In the last weeks of 1997, case in point, *ABC Nightly News* ran a story about shopping on the Internet, the upshot being that, despite the concerted efforts of big money, not much of it was going on. Apparently, it doesn't matter how often people are told that shopping on-line is no riskier than handing a credit card to a waiter. They still prefer to do their shopping on-premises.

Of course, this being America, no expense will be spared in trying to turn the Internet into a commercial bonanza—one would have to be willfully myopic to think otherwise. But wanting and having are two different things, and when it comes to online trade, some serious obstacles remain stubbornly in place, not least of which is that trust has a basis in biology. In fact, when you start to look at the problem this way, a number of seemingly unrelated subjects snap into sudden focus.

One of the best books ever written about trust is *The Selfish Gene*, by Richard Dawkins. That *TSG* is ostensibly about population genetics matters not a whit. Pick it up sometime and you'll find more than your fair share of insights into the vagaries of human behavior.

Front and center, Dawkins proposes that the unit of evolution is not the individual, but the gene, because the gene is the thing that perpetuates itself through generations. This explains why you'll be quicker to save your own kin than someone of another race: the genetic program thinks of itself first. Nevertheless, says Dawkins, there are times when it behooves genes to drop their survival shtick and do some collaborating. On one level, this cooperative habit is clear enough: since time immemorial, genes have banded together to make up the genomes that reside in us all. But genes will sometimes work together in a less obvious manner: namely, when their "host" organisms step out of character and decide to cooperate among themselves.

To illustrate this point, Dawkins gives the example of a kind of bird prone to getting ticks on its head. Because of obvious limitations, no bird can ever hope to get a tick off its own head; to do so, it needs the assistance of another bird. So how do these birds react to their situation? Well, set a flock of them down on an island somewhere and at first they'll play nice, removing each other's ticks in kind. Then, eventually, a few birds will discover that they need not return the favor: they'll become Cheats, and their gullible counterparts will end up as Suckers.

Now, a population made up of Cheats and Suckers is already a pretty interesting thing, but the game becomes more interesting again when the organisms in question can recognize individuals and judge them according to past performance. This ability can be either general or specific depending on the species. Crickets, for example, have a general memory, in that they can only recall their *own* past performance. If a cricket loses a few fights in a row, it'll start to cry uncle in the face of any aggressor. Monkeys, on the other hand, have a specific memory, which allows them to distinguish one individual from another. Blessed with these keen perceptions, they tend to become docile only when they meet up with monkeys who have bested them before.

Understandably, possession of a specific memory will change the rules for the tick-blighted birds as well: if a stranger refuses to pick the tick off a bird who happens to have a specific memory, you can be sure the favor will go unrequited next time. Dawkins calls the creature that exhibits this kind of behavior a Grudger. The Grudger will behave exactly like a Sucker . . . until he gets burned. And since of the three types—Cheat, Sucker and Grudger—the Grudger has the best chance of survival, evolution can be seen to favor a certain amount of trust.

So do people go around cheating and sucking and grudging for the same reasons animals do? Dawkins seems to think so. "A long memory and a capacity for individual recognition," he writes, "are well developed in man. We might therefore expect reciprocal altruism to have played an important part in human evolution."

This last point seems obvious enough from what we know—one need look no farther than the shade of the nearest Christmas tree. What's more, throughout history, most human exchanges have taken place in settings where both parties could recognize and remember each other. Atrocities have abounded, of course, but the perpetrators as a rule have been easy to spot (think of Napoleon's soldiers, displaying their colors as they advanced), and the tradition of the human Grudger has prospered—or at least taken hold—accordingly.

Things began to change a mite, however, as the Industrial Revolution came to a close. You could point to the ability of executives to hide behind large corporations, but the rise of the first great department stores in the late 19th century is just as telling. Consider, for example, this passage from William Leach's *Land of Desire:* "Before 1880, most people bought raw materials in bulk and carried the purchases home themselves; there were no packaged goods and relatively little ready-to-wear clothing. Customers and owners often got to know one another very well; and, if the contacts were good ones, it was not uncommon for merchants to offer services to customers over long periods of time . . . Between 1880 and 1915, the new economic conditions undercut this face-to-face interaction. Merchants began to sell at fixed prices as manufactured, packaged, and ready-to-wear goods multiplied. So many ready-to-wear articles proved defective, so many items often crumbled or were soiled in transport and delivery, and the trade volume was so large, that the older grounds for trust between customers and merchants had been undermined and new grounds were required."

This is a near-perfect illustration of the Dawkins scenario with the recognition factor removed—a crisis, you might think, a brief foray from which all sensible parties would quickly withdraw. Ah, well. To silence accusations of defective goods, store owners initiated a return policy, which prompted the now-familiar strategy of buying with a premeditated intent to return. To keep the large volumes of stock flowing out the door, owners upped the ante with an amenity called the charge card, which led to rampant abuse on the part of buyers and eventually—naturally—to today's ubiquitous credit card.

All of this must have been exceedingly strange for everyone involved. It wasn't so much that people were discovering better disguises, it was more that the act of cheating was being dispersed and displaced without actually being eliminated.

Of course, face time has only decreased since then. From the department store, we've progressed to the home shopping channel and, *mutatis mutandis,* to the prospect of the on-line bazaar. It would be tough going to find the soul who believes that trust has broken out all over during that time, yet oddly enough, proponents of Internet shopping often use the notion of incremental change as their central defense. (Actually, the standard argument is that commerce over the Internet is no different from giving your credit card number over the phone. But surely these two cases are not exactly the same. After all, if they're *no different,* then why go with the new system at all?)

We know what shopping over the modem means to industry insiders: it means sales upon sales, 24-7, without anyone to scuff up their welcome mats. Unfortunately, a Dawkinsian reading offers a good deal less pep. The cricket, remember, can only recall his own past performance and will turn docile after several defeats. It seems clear enough that the same thing is happening to Americans today—unable to locate, much less resist, the cheat in the machine, we're becoming little more than pliable tourists. We're becoming the crickets.

Except, perhaps, for a few assorted malcontents. After all, it stands to reason that a small portion of Grudgers, if repeatedly denied the ability to ID the Cheats, will start to lash out at random. By the same count, it can hardly be a coincidence that terrorism was born in the very same era that saw the birth of the modern corporation and the department store.

Consider the perspective of an average brass-worker in 1883, as quoted by Henry David in *The History of the Haymarket Affair*.

"I remember that fourteen years ago the workmen and the foremen and the boss were all as one happy family; it was just as easy to speak to the boss as anyone else, but now the boss is superior, and the men all go to the foremen; but we would not think of looking the foreman in the face now anymore than we would the boss . . . The average hand growing up in the shop now would not think of speaking to the boss, would not presume to recognize him, nor the boss would not recognize him, either."

This was not simply a matter of turning one's head when an enemy passed. As labor leader P. J. McGuire described it, employers and workers were becoming actual strangers to each other: "They do not know each other on the street."

The effects of the new estrangement made themselves felt soon enough. In 1887, Andrew Carnegie, safely situated far from the huddled masses, challenged his fellow robber barons "to show that there is pauperism" in the United States. William Graham Sumner went further, saying, "It is constantly alleged in vague and declamatory terms that artisans and unskilled laborers are in distress and misery, or are under oppression. No facts to bear out these assertions are offered."

When you and your boss are no longer on speaking terms, it's pretty hard to be anything but vague and declamatory. And if being vague and declamatory only leads to your boss being equally vague and declamatory, you might eventually do what an unknown stranger did in a Chicago crowd on May 4, 1886—despair of all dialogue and throw a bomb at the cops. That incident, which sparked the Haymarket Riot, was the first major act of terrorism in the United States and a defining moment in the labor movement. It also fits Dawkins's description of a fourth behavioral type—the Retaliator—with frightening accuracy.

I suppose a case could be made that the facelessness of Internet shopping will shatter the ethnic favoritism endemic to commerce since there was such a thing; from one point of view, the on-line market looks like a fulfillment of the multiculti dream in spades. But it also seems fair to point out that the same trend is sure to create some new wave of disaffected Grudgers who don't feel all that particular about their aim. When the levee breaks, everyone's bound to get wet.

January 7–13, 1998

The War of the Web: Round Two

IN THE SUBURBAN neighborhood where my mother lives, the passing of decades has left the streets as implacable as ever, the lawns as timeless as cheese-food product. The crime rate, you could say, has jumped from zero to nonexistent. Yet something weird is going on. In the past year or two, cars have begun to line up outside the grade-school doors, manned by parents too nervous to let their children walk home.

What exactly are the citizens of America sensing out there in the woods? An imminent UN takeover? Hordes of blind Muslim clerics?

A friend of mine suggests that it's a case of insular communities watching too much violence on the news—that today's commuters are quietly clogging their brains with outsized visions of mayhem. This explanation doesn't seem exactly right, though, because if anything the news has only grown tamer of late. I mean, who can recall a time when the weather, with its blistering blizzards of blarney, has topped the hour so often?

The *intimation* of violence, on the other hand, is another story. In 1995, TV programmers stepped up their reports of BBS-Nazis, bombmaking recipes and, when those fail, e-mail obscenities. To believe that any of these developments could result in an afterschool abduction, you would have to imagine a lurker in possession of his own satellite, zeroing in on the coordinates of a bob tail bouncing in the wind. But that's just the thing. Internet scare stories don't sate you with actual details. They keep your fear nodes palpitating. And so, as V-chips jam the assembly-line gates and the

German government blocks on-line services (which it did again last week, blocking the Web site of one Ernst Zundel, a neo-Nazi), cyberspace is fast shaping up as the Charybdis of the '90s, jaws open at the horizon, waiting to swallow wayward travelers.

By now, the debate has settled into its middle phase, with cogent points being raised, confusion abounding and the mainstream media waiting for another roil of the waves. Normally, I'd be content to let the whole fiasco pass, but as one who learned everything he needed to know in the 19th century, I feel honor-bound to point out that all this has happened before. And since 1996 marked the centennial of Guglielmo Marconi's first radio patent, it seems especially fitting to track the fate of wireless as it relates to the Web.

Marconi, like the designers of the Internet, couched his dreams in military terms from the very beginning. (The Internet was designed to withstand a nuclear blast.) Finding an able patron in scientist-inventor Sir William Preece, he left his native Italy for England in 1896 and, on September 2 of that year, began inviting Army and Navy observers to his experiments on Salisbury Plain. The armed services were never far from his side after that. When he organized the Wireless Telegraph and Signal Company in 1897, it was to service ship-to-shore communication. "I believe one of the greatest uses to which these instruments will be put," he proclaimed, "will be signaling in wartime."

But just as the Internet fell into the hands of hackers, so too did radio slip into private hands. By 1910, the Wireless Club of America could boast 10,000 amateur operators as members, and by 1912, there were an estimated 122 wireless clubs in the U.S. alone. These amateurs invariably spoke of their hobby with the same fervor that propellerheads reserve for computers today. "Imagine a gigantic spider's web," wrote Francis A. Collins in an uncanny choice of metaphor, "with innumerable threads radiating from New York more than a thousand miles over land and sea in all directions . . . Our operator may be compared to the spider, sleepless, vigilant, ever watching for the faintest tremor from the farthest corner of his invisible fabric."

In fact, the amateur operator resembled a hacker in almost every way. Typically a white boy without gymnastic portfolio, he was tinkerer, inventor and nerd-king rolled into one, operating just outside the niceties of social consensus. He transmitted and received, point-to-point, limited only by the reach of his equipment. Though voice transmissions became feasible after 1906, he often stuck with the older Morse-code methods—much as a contemporary hacker will favor code over icons. Airspace remained wide open, and there was no FCC.

Sadly, the frontier days were short-lived. As the ether grew thick with chatter, the Navy began accusing amateurs of taking up valuable airspace with gossip, spreading false information and—flamers, take note—transmitting obscene messages. In another complaint that should resonate in the modern ear, the amateur could wreak havoc without being detected, which branded him as a threat to national security. The amateurs, for their part, claimed that inept Navy operators were trying to cover for their own mistakes.

The situation came to a head in 1912, when the *Titanic* met its spectacular doom. With emotions running high, it was quickly asserted that amateurs had jammed the lines and slowed the rescue mission. It was also asserted that Marconi's operator at the New York Wannamaker store, David Sarnoff, had been the only operator at the key throughout the catastrophe.

Neither of these assertions was true. The distress signal was first received by a Marconi operator aboard the *Carpathia*, and relayed from there to Cape Race. Two other ships sailing closer to the *Titanic* suffered from inadequate wireless equipment and so missed the signal altogether. As for Sarnoff, he was present at the Wannamaker station to relay some of the news, but other Marconi stations continued on after his was closed.

No matter. Seeing a monopoly in the making, Marconi himself leapt to the fore, recommending "control over amateur experimenters." The Radio Act of 1912, which grew out of the momentary fervor, gave the lion's share of the spectrum over to the Navy, pushing amateurs into the short waves. During the war, the Navy gained control over many radio patents, which, when peace was again restored, were appropriated by that vigilant boy at the key, David Sarnoff. Now grown to be a man, Sarnoff formed RCA and turned radio into a fixed, one-way phenomenon: the broadcast medium familiar to citizens of the late 20th century.

So much for the War of the Web, round one.

Of course, radio transmissions and e-mail are not the same thing. No amount of on-line use is going to prevent an emergency bulletin from reaching the Coast Guard anytime soon. Radio airspace is subject to the laws of scarcity; for all intents and purposes, the Internet is not. But none of this really matters, because amateur operators never interfered during the *Titanic* disaster in the first place. In fact, the great lesson of the *Titanic* (besides as a caution against the use of adjectives like "unsinkable") is that there's nothing like a catastrophe for regulating new technologies. In the present scenario, all it would take is the introduction of digital cash, followed by an on-line heist or two, and the rights of on-line users could be summarily revoked. But hey—at least the kids would be able to walk home from school.

February 7–13, 1996

The Handwriting Is on the Screen

AT FIRST GLANCE, Kate Gladstone doesn't look like an activist. Her PTA-style dress, her oversize plastic glasses and her well-nourished face suggest a more suburban character—a den mother, say, or a home-shopping enthusiast. Yet "activist" is a label she would probably relish.

"I'd like to see the picketing of schools," she announces, brandishing a large, plastic-carrot pen. "Maybe kids going on strike. I would even consider a cropduster plane that dropped leaflets containing inflammatory remarks."

Is Gladstone primed to do battle with Operation Rescue? The porn industry? The right to bear arms? No. Her juggernaut is aimed at the *handwriting* establishment.

Gladstone probably knows more about penmanship —its history, its argot, its politics—than anyone else alive. She speaks of it with a fervor generally reserved for proselytizers, pulling out arcane facts, summoning vast stores of statistical knowledge, occasionally raising her voice and her person in frenzied appeal. And she's just as willing to go to the barricades for her Patent #5,018,208, which covers a new kind of handwriting ID.

Imagine a pen that's connected to a computer. Several sensors are embedded in the grip of this pen, down by the nib. When you start writing, these sensors detect the amount of pressure you use from moment to moment, including the time when the pen is in the air. Combine this pen with one of the technologies that can write on a computer screen and you've got yourself a powerful identification unit. Such a pen, in recognizing both tactile and graphic trademarks, would rival the accuracy of a DNA test.

"There are pressure-sensitive pens around," Gladstone says, articulating her consonants at lightning speed. "Mine is not only different but better. Most pressure-sensitive pens have their sensitivity in the point, which could be damaged. My sensors are in the grip, which means that they can be replaced if they get damaged or worn."

Obviously, this kind of pen could find many applications in work situations. A drunken worker who had to sign one of these contraptions would be hard put to simulate a sober signature. Similarly, forgeries would be difficult to achieve because, as Gladstone puts it, "a forger has the choice of doing it right or doing it fast." Conversely, honest deviations from a signature would pose no problem: just have the worker sign a number of times, preferably during the initial job interview, and store the variations in the database.

Against Big Brother charges, Gladstone defends her invention as being less cavalier with constitutional rights than the usual drug-testing procedures. After all, she points out, signatures measure performance, not party behavior. Less defensible are possible abuses by graphologists, or handwriting analysts (although locking up people who draw circles over their "i"s might not be such a bad idea).

But Gladstone doesn't seem worried about such things anyway. Instead, she prefers to conjure wider applications: a car designed for DWI offenders that requires a robust John Hancock before it will start, handwriting samples for ATM machines or, more dramatically, high-security clearances.

"Let's say I work in a nuclear power plant," she muses. "What's to stop some terrorist from coming in, knocking me unconscious, chopping off my hand, sticking it against the fingerprint pad, sticking my eye against the retinal ID, taking out the chip in my forehead, heating it with a hair dryer by hand until it stays at 98.6?"

Handwriting, it seems, can lead a body down some fairly exotic roads. But this, I learn, is nothing. Having dispensed with the future, Gladstone now sets her sights on the past.

"Why do school children learn first to print and then to write in cursive?" she asks rhetorically, then breaks into full epic stride. What she delivers is not the usual liturgy on the Phoenicians and the Babylonians and the rest, but the unkempt tale that the history books left out.

In the Renaissance, Gladstone says, scholars devised the first version of the italic (as in "Italian") alphabet. At that time, italic used a combination of joined and unjoined letters, depending on expediency. This style went out of fashion during the Baroque period, when the invention of copper plating encouraged smooth, continuous lines, and thus an ornate script that showed off the talents of the engraver. From the Baroque style developed the florid penmanship taught in grade schools as late as the early 20th century.

In recent years, however, as the pace of life accelerated, classes in ornate handwriting came to be seen as time-consuming, so an alternative was sought. Marjorie Wise, an Englishwoman, thought she had the answer when she brought a style composed of verticals and circles to Columbia University. Her system filtered into public schools in the '40s and '50s, and might have continued unchecked had some people not noticed how difficult it was to master. In classic bureaucratic style, a compromise was struck. The upper grades would keep the old cursive system, while the youngest grades would learn the circles-and-sticks version, which most people refer to as "print." As a result, two disagreeable systems prevailed instead of one. Meanwhile, Wise went back to England and recanted her position. Verticals and circles *were* too difficult, she decided. Too late, though: her system is still taught to unquestioning students today.

Gladstone cares about all this because when she was growing up she had dysgraphia—an inordinate difficulty in writing. Not content to take her disability lying down, however, she went the other way and became a heroic penwoman. She even went so far as to develop her own system of handwriting based on the old italic system; with letters that are joined or not, depending on which course is easier.

Making handwriting easy is important, Gladstone says, because "the increase in vocabulary can be linked to performance in handwriting." There is some sense to this. After all, if you can't write "Tyrannosaurus rex" without getting a monster cramp, you're likely to settle for writing about dogs and cats and trees and rocks, as Gladstone did when she was young. Literacy may depend on legibility more than we know.

And here's where the activism comes in. When Gladstone wrote to the CEO of a major handwriting education company, telling him of her italic redux, he wrote back challenging her to a handwriting duel. Gladstone accepted, only to be given the cold shoulder. It was undignified, came the response. Every other attempt to connect with the establishment has met with the same results. Apparently, the powers-that-be are content to perpetuate the existing situation.

Her path to legitimacy effectively barred, Gladstone now intends not only to commercialize her pen but to hang out a shingle as a "handwriting repairperson." "I'd like to have SWAT teams—for Special Writing Assistance Teams," says Gladstone. "But of course, if I recommended that my system be taught in public schools, that same company would firebomb our house."

Again she begins to bubble over, her voice rising, her excitement running away from her. For a second, I wonder whether she's crossed over the line. But then I remember—if she had to take a handwriting test to determine her competency, she would pass. Every time.

February 13–19, 1995

5

IN THE REALM OF THE SENSES

Visions You Can Have at Home

YOU PROBABLY KNOW me well enough by now to expect that I will follow any lead, no matter how dubious it may be. Well, once again, I have lived up to the demands of my exacting public. The other day, I was wandering around Tower Books (which is, let's face it, essentially the Internet for the phoneless), when a small-press book attracted my eye. The cover stated simply: *Dreamachine Plans, created by Brion Gysin.* Inside were diagrams, templates and a short text on the virtues of inward excursions.

Shades of *The Phantom Tollbooth,* I thought, recalling Norton Juster's fairy tale about a mail-order device that allows entry into other worlds. But the byline also made me think beyond sentimental journeys. For those who can't place the name, Gysin developed the literary technique of cut-ups, thus sending William Burroughs on his way. He also introduced the Joujoukan musicians of Morocco to the West—through the Rolling Stones, no less.

I always had a lot of respect for Gysin, and I maintained an interest in his activities over the years. Apparently, the Dutch company Phillips did, too, because for a while they entertained the idea of patenting the Dreamachine. Nothing came of it, of course, but that isn't so surprising. There's little evidence that Gysin knew how to do lunch.

Be that as it may, the Dreamachine—alternatively called a flicker machine—is a neat piece of work. You cut a specific array of holes in a square yard of cardboard, which you then bend into a big cylinder. Place the cylinder on a 78 rpm turntable, suspend an ordinary light bulb down inside and set the gizmo spinning. Pretty simple stuff, really—a combination magic lantern/strobe light. In fact, things get counterintuitive only when the viewing method comes into play. In a text that's generally reserved in its claims, this detail is trumpeted in declamatory style: "ONLY ONE OBJECT HAS BEEN MADE TO BE VIEWED WITH THE EYES CLOSED: THE DREAMACHINE."

The payoff? "The development of autonomous 'movies,' intensely pleasurable and, possibly, instructive to the viewer." The flickering light—so goes the claim—creates complex, abstract patterns on the back of your eyelids, eventually giving way to a deeper state, where representational imagery appears.

All right, I thought. I can do this. No problem.

Would that it were so easy. It helped that I already had a Victrola. But by the time I had traced and cut the required 40 holes, I began to have an inkling of what real inventors go through. The backside of my Exacto knife had destroyed the nerve endings in the tips of both index fingers, and the cylinder sagged like a glove in the rain. Only after some elaborate jerryrigging did my flicker machine stand up straight—a sad sack of a prototype, to be sure, but workable nonetheless.

I turned on the light bulb, started the Victrola and shut my eyes.

The initial effect was alarming. A bright kaleidoscope of color flooded my view, and my immediate thought was for my health. (Indeed, Gysin warns epileptics not to try this at home.) Gradually, however, the assault turned into a manageable dazzle. I began to see whirling rainbow patterns that followed an ever-changing series of orbits, as if all the rides at Coney Island had been filmed at night and then strung together in a sequence of dissolves. Other patterns resembled the Northern Lights—shimmers and starbursts in shifting fields of color. Moving my head in relation to the cylinder, I discovered one area where a soft orange light pulsed slowly and intensely. This was the sweet spot, the mother of all flicker patterns, and I cottoned to it like a monkey in love with a stick doll. When I finally turned the flicker machine off (not overly aware of how much time had elapsed), I felt a curious well-being, as if my brain had taken a bath.

All this, mind you, from a guy who hasn't ingested a psychedelic drug in fully two decades. After a while, I was able to predict the course of my flicker sessions—to an extent. I could get to the whirling rainbows whenever I pleased, for example. But other effects came as a surprise. Layman that I am, one experience even hinted that a flicker machine could have medical value.

I have an astigmatism, which means that my left eye receives the world in a disorganized fashion. What would happen, I wondered, if I were to "look" at the flicker machine with just my astigmatic eye?

I found the pulsing orange light again and slowly turned my head until my right eye was out of range. At a certain angle, the light seemed to freeze behind my left cornea. The circulation in my sinuses improved noticeably, and the muscles in my left eye, long dormant and snug in their socket, began to jitter and tremble, as if coming alive. Frankenstein's monster never had it so good.

Alas, a fit and trim left eye was not in the cards—not after four days anyway. I did experience other notable results, however. After the typical flicker session (about 10 to 15 minutes long), I felt a pleasant warmth behind my eyes. If I worked on my computer, I tended to register split-second changes on the screen more acutely. Strangest of all, other people's eyes began to look unusual—like two distinct features on a single face. This was puzzling until I realized that when I look at a person's face, usually I focus on the left eye. Now, for whatever reason, I was looking into the right eye. It was as if I were talking to a slightly different person.

Clearly, the flicker machine had me in its thrall. During idle moments in my day, I could hardly resist turning it on. I was hypnotized—and, I guess, a little deluded. When I explained my adventure to friends, they patted me on the hand and deferred with peevish smiles. One of them delicately pointed out that life itself is made up of incredible light patterns.

Not an easy point to argue, unless you consider that television executes its own light patterns to great hypnotic advantage and, as it turns out, is also a distant cousin to the flicker machine. Back in the 1920s, the early TVs incorporated a device called a Nipkow disk, a whirling thing with a series of holes that created a continuous image when light was passed through. Then, too, during the same era, an inventor named Hugo Gernsback promoted a TV set that, like a flicker machine, could be built at home. The inventor of TV himself, John Logie Baird, was hardly a corporate type. He embarked on his mission to broadcast images only after failing at ventures in such urgently needed goods as paper socks and glass razors (no kidding). I can imagine him and Gysin getting along famously.

Granted, I never saw the pseudo-events and fabulous landscapes that others claim to have seen. But these days, with cyberspace fast becoming the plaything of the moneylenders and techno-groovy relaxation machines coming over the transom, anything under $20 starts to sound pretty good. And I know that I'm dreaming deeply now, because the world is looking brighter than usual, but wouldn't it be a pleasant surprise if someone were to rethink the TV set from scratch, and this time it actually had something to do with vision?

May 4 – 10, 1994

Truth in Mirrors

JOHN WALTER doesn't want much. He just wants you to see yourself as you never have before.

"We'll get some reactions a little later," he says, his True Mirror as yet unveiled on the chair beside us in the usual crowded cafe, "because it ends up being a kind of litmus test for how people are."

The inspiration came to him about 10 years ago. Seems he was having a grand old time at a party until he looked into the medicine-cabinet mirror and saw something unpleasant, something foreboding. When he turned one of the mirrors at an angle, however, he got that "non-reversed" effect known to self-oglers the world over, and his buoyant mood instantly returned.

From that day on, Walter wondered how he might construct a mirror that would give a non-reversed reflection of the human face. The idea sat on his shoulder while he tested satellites to see if they could withstand nuclear war. (The job ultimately disgusted him.) It whispered in his ear while he underwent studies at Life Springs, an organization he describes as a cousin to EST. ("It's too bad they started proselytizing.") Then, about two years ago, he decided that he was simply going to make this True Mirror thing work.

A cinch, you might say, but in practice it's not so easy. Take two mirrors and place them at a right angle and you're sure to be distracted by an unsightly seam running down the center of your eyes. Walter thought that was distracting, too, so he struggled to render the seam invisible.

It turned out that most of his problems had already been solved in the piecemeal advance of history. About

100 years ago, a man lost to history had the bright idea of putting the silver used to make glass reflective on the front of it. (Since we're talking about a three-dimensional physics problem, you might want to take my word for it when I say this makes a difference in the True Mirror experience.) Of course, silver tends to scratch at the merest nick, but here Walter lucked out again, because in the century-long interim, another sharp thinker had invented a polymer no more than three atoms thick. With this winsome plastic, Walter could coat away to his heart's content.

Next was the problem of mass-producing a mirror with an extremely straight edge, so as to keep the seam tight. This was not tidily accomplished either, since stress fractures tend to have small jags on them. But by using glass of the right width, Walter soon had an edge true enough for government work.

The mechanics thus explained, I play devil's advocate. People can see themselves on videotape easily enough, I suggest. Wouldn't a camcorder fulfill the same function?

"Videotape doesn't give you real-time representation," Walter replies, maintaining eye contact with an unnerving candor. "It shows you a version of yourself that's already gone. A mirror, on the other hand, provides a feedback loop, so you can change your image while you see it." As for those video cameras in store windows, Walter rightly points out that real-time video images don't let you see yourself looking into your own eyes. Your gaze is inevitably trained away from the monitor.

Suitably convinced that Walter has thought this thing through, I suggest that it's time for a look.

I hadn't expected much, I guess. I had read the press release and noted that about 10 percent of the people surveyed saw something important about themselves in the True Mirror. But, you know, about 10 percent of the people *I* surveyed think a Chia Pet adds tone to any home.

Well, what the hell.

What I see is a little disconcerting. I look into what should be two mirrors set at right angles, but it doesn't seem like two mirrors. Walter has done quite a job of getting rid of the seam, all right. There's my face in the center, crooked smile to the other side. I have the sensation that my face is floating somehow. It's pleasant and disorienting at the same time.

Not much of a scientific test, I'll admit. It happens to be a beautiful day, and I happen to be in my first good mood in weeks. I'm also not the kind of guy who responds well to solicitation. Ask me if I like a movie before the house lights are on and I might pan your mother. So when Walter wants to know how his mirror makes me feel, I feel my jaw taking on the shape of a clamshell—and shutting.

All of which only serves to demonstrate a certain point: mirrors are so damn psychological. Are you the sort of person who will steal away to peer longingly at your mug until it changes shape, changes age and starts to run numbers on your soul? Or are you the type who can't bear to see yourself—an irritable, naysaying Narcissus? Either way, there's no way around it: mirrors and humans make intense partners.

"When I first looked in this mirror," Walter tells me, "the message I got back was that I was really okay. I think a lot of people, when they look in a regular mirror, start to feel bad about themselves. But what I've discovered in showing this around is that people who have some idea about themselves will really respond to it."

As if on cue, someone from the next table over approaches us and wants to know what the buzz is all about. He looks into the True Mirror with a dawning fascination, takes a quick snapshot and segues right into business.

"Right now, they cost $150," Walter asserts. "When I get into bigger production, they'll be about $100."

Another guy comes over and starts asking questions. He's a little more tentative, but cheerful nevertheless. He looks furtively at his reflection, then speculates about possible applications for it: in clothing stores, on movie sets.

Though his background is in physics, Walter couldn't care less about scientific applications. "If they use it, fine," he says. "I'm more interested in people."

And he is interested in some fairly interesting people at that. Once, for example, he took a prototype to one of the famous Rainbow Gatherings, where the majority of his subjects, it can be assumed, were under the influence of some kind of powerful psychedelic drug. Not your usual looking-glass-lovers, but lo and behold, the True Mirror was a hit. A whopping 75 percent responded favorably.

Indeed, one can only guess what effect the True Mirror might have on the many unsolved personalities of the world: schizophrenics, manic-depressives, autistic children. For the nonce, however, Walter is happy enough to have forged a mirror worthy of the production line. As he polishes his prototype and starts to wrap it up, he relishes in its sturdiness for a moment, pointing out the fine points of its design.

And if it should fall, perchance to break?

"Seven years," he says, laughing. "Of good luck."

March 30 – April 5, 1994

How the Brain Sees It

EVERY NOW AND THEN, it pays to take a break from inventions and look at the thing that's actually doing the inventing. I'm speaking, of course, of the glorious man-of-war that bobs inside the skull, that complicated cauliflower—the brain.

Few subjects offer less in the way of neat explanation than neurology. The brain has often been compared to a machine, its many activities tooled to the finest precision and timed with a clockwork majesty. Yet much of what goes on up there, upon closer inspection, seems anything but elegant. Try this experiment, for example. Ask a friend to name 10 vegetables, then watch his face as he reels off the list. Chances are, his eyes will dart all over the place, as if he's looking for the answers in the corners of the room. Nothing in those corners will give him the answer, yet off his eyes go, this way and that. If the brain is so lean and mean, why does it waste its time with this bit of stage business?

That was the question put to me by Dr. Keith Purpura, an assistant professor in the Department of Neurology and Neuroscience at the Cornell University Medical Center (and, as it happens, a friend from high school). The answer, or at least the first sally toward one, is contained in a paper with the formidable title *The Thalamic Intralaminar Nuclei: A Role in Visual Awareness*, by Purpura and Dr. Nicholas Schiff, a clinical neurologist, in the January 1997 issue of *The Neuroscientist* (http://www.theneuroscientist.com). In fact, though it's premature to be blowing any trumpets, this paper may hold the key to far more than a simple parlor trick. Ultimately, it may provide insights into the nature of dreams, memory and a whole host of pathologies.

"I brought these," says Purpura as we take up our positions in a restaurant awash with models and manques. "These are representations of different parts of the brain, drawn by a man named Wendell J.S. Krieg. But they're not like the usual diagrams of the brain. They're drawn as if you were inside looking up at the surface."

I look down at a stack of drawings that resemble something out of a Jules Verne novel. Slender ganglia of tubes crisscross dark, billowing folds, rendered throughout in the texture of rat tails. I get the distinct impression that this Krieg fellow went spooky somewhere along the line. But apparently he also anticipated Purpura's needs quite well, because these drawings will serve as our map as we embark on the road that leads from life to the life of the mind.

Of course, science being what it is, the road is a fairly windy one, and we have to start with some basics. As you may recall, most perceptions are received in the back of the brain, while judgments about those perceptions are made in the frontal lobe. When you see a car coming, you're using the back of your brain. When you decide to step away from its path, the front of your brain is at work. Obviously, this setup requires information to move from the back to the front—and, one hopes, to arrive before it's outdated. But how it gets from here to there is another matter.

While much of this brain-wave migration remains a mystery, a few observations have been wrested from the darkness. Certainly, a good deal of processing takes place en route from aft to fore, as raw perceptions are molded into intelligible information. It's also known that waves of activity will compete with each other, with one dominating or even blocking out another. Scientists have even observed secondary waves that reflect backward—an undertow, you might say—as the main waves propagate forward.

And then there are those eye movements. "Most states of attention last no longer than one to two seconds," says Purpura. "But within them, lots of other things are happening. For example, the way you look at the world, your eye moves all over the place. It zigzags. Actually, what it's doing is moving fast, stopping, moving fast, stopping."

These stopping points, each of which lasts about 200 milliseconds, can be said to constitute basic units of thought, because they impose a time limit during which the brain must do its work. An apt comparison is that of an emergency room. Patients are coming in all the time, and they have to be sized up and sent out fast, because a new crop is already on its way in. Awareness, it turns out, is a lot like triage.

And so we move to the major premise, which says that if the brain is such a beleaguered master, it has to be able to coordinate perceptions and actions. This makes plenty of sense when you consider the alternative. Say the phone rings. You pick up the receiver and hear someone's voice, which prompts you to say something back. All of these events have to be synchronized, or you'll end up talking into the flower pot.

What Purpura and Schiff are banking on is that the thalamus, an almond-shaped area located deep in the centei of the brain, plays an important role in this coordinating process. To be more exact, their quarry is a very small portion of the thalamus called the intralaminar nuclei, or ILN. For bursts of about 200 milliseconds, Purpura believes, the ILN operates like an air traffic controller, facilitating the information pass from the back of the brain to the front. Supporting this hypothesis is the data that shows the thalamus sending out bursts of activity at the very moment when the eye comes to rest.

It's a compelling scheme. The eye wanders for a moment, then stops, giving the go-ahead for the ILN to send out the master code that tames the electrical storms inside the brain. Then the ILN stops and the eye wanders anew.

"We're trying to find what's wrong with this picture," Purpura says. "We would like to know. But everything we've seen so far only strengthens it."

As a scientist, Purpura is not really in the business of saying where this knowledge will lead (should it pan out), but he has some ideas nonetheless. For one thing, fast eye movements, like REM or those seen when some people try to recall information, may reflect participation of the ILN. On another level, the ILN makes its signature burst-feed patterns not only when the eye scans a room but also during the process of waking up—suggesting multiple purposes for the mighty thalamus. But it may be in the field of medicine that the most valuable applications are found. It's conceivable, for example, that 21st-century stroke victims could be helped by some sort of action on the thalamus.

In the great debate over the nature of consciousness, Purpura comes down squarely on the side of materialism, which says that awareness springs from biology and biology alone. As a materialist, he gives no truck to spirits inhabiting this earthly vessel, or anything like that. Still, spending a few hours thinking about thinking can be an oddly metaphysical experience.

I left Purpura that night with the disconcerting suspicion that the scene unfolding before me was not quite as tangible as it should have been. A garbage truck grumbled by. Two women in furs were swept down the sidewalk on the jet stream of their own perfume. A beer can gleamed in the cheap and noisy lights. But I knew now that all of this was just a reflection. Somewhere in the middle of my head, fireworks were going off.

January 15–21, 1997

Levitation, with Sirens

YOU WOULD NEVER know from the looks of his live-in studio that Dr. David Deak has an inventor's mind. The space is airy, with parquet floors, a spotless efficiency kitchen—more like a middle-class way station than the site of bold new experiments. True, there is that Plexiglas cube on the table, and a chalkboard clouded with thetas and gammas. But seeing is only a small part of the formula here. To reach the place where Deak does his deepest dreaming, you must broach the field of acoustics.

Perhaps you remember that old liner-note chestnut, for example—the one about sound and its power to uplift? Well, Deak has turned it into fact.

"It occurred to me while I was reading about the air-raid sirens used in World War II," he says in his undiluted British accent, his beard brought low to his barrel-thick chest. "I had some identification with that, because I was born at the height of the bombing—1942."

In those days, air-raid sirens were literally loud enough to peel paint, which was not a good thing, because the paint chips tended to clog up the horn. "They thought the chips were bounding around," explains Deak. "I thought about that. If the chips were actually floating, and you could keep the energy pulsing, then they should continue to float."

We step over to the Plexiglas cube (known as a Helmholtz resonant cavity to friends) for a little demonstration. Inside it lie a few bits of Styrofoam squiggles, maybe 15 of them. Deak mans his computer, which controls the sound he pipes into the cube.

Did I say sound? A cauterizing screech is more like it, a caterwaul, or maybe—well—an air-raid siren. Not exactly talk-show material, unless Marlee Matlin becomes a late-night host any time soon. But sure enough, while my ears are adjusting to the blast, a few of the Styrofoam bits rise up and hover, dead center in the cube. Deak makes them twirl, a pirouette in the name of physics, then keeps them stable again. They dangle there until the screech abates.

Like a good constable, I inspect the specimen. It is plain that wind has not been used, nor suction, nor magnetic devices. In sum, your lordship, the demonstration is correct. Sound can levitate objects.

Deak leaps to his chalkboard with the ease of a seasoned professor—he taught at the University of Pittsburgh—and exhales a steady stream of physics. The levitation works, he explains, because the sounds in the cube form standing waves. And since sound waves create pressure, a standing wave will create a set of stable pressure zones. Following these principles, Deak has set up a ground zero, if you will, of high pressure at the center of the cube toward which the objects "gravitate."

"And so you see," concludes Deak, "it is, for all intents and purposes, an anti-gravity machine."

No doubt the wags in the back row will be scoffing by now. Even if it's true, they will sneer, do we really need another plaything to distract the kids while the squadrons strafe the city? Or at best, another misguided dream of Hovercrafts to whisk us off to yesterday's tomorrow?

All good points, but Deak has other uses in mind, most of them a bit more mundane than that. Although he is laboring with Churchillian technology at the moment, his plan is to move on to ultrasound, which can handle denser objects at considerably lower decibel levels. If lenses could be held aloft for cutting, for example, they might be spared the impurities that come from graphite molds.

(And you may recall the tale of the Hubble telescope and all that bothersome dirt.) The device might also be used to stratify powders, or to levitate anything else that isn't magnetic.

"But it's really up to others to come up with applications," says Deak with a tinge of pride at his patent-pending device. "I have made visible the phenomenon."

The acoustic levitation chamber, besides being a phenomenon, may well be Deak's Rosetta Stone: it has spurred him onward—and, you might say, upward—to other inventions. Going on the principle that sound waves will move fluids, for instance, he recently enlisted a machinist to make a parabolically tapered tube. Put an ultrasonic transducer on the large end, bring liquid in through the side and you've got the makings of an acoustic pump with no moving parts.

In theory, he says, any kind of pump could be driven by sound, including an artificial heart. "It would have to be ultrasound. You wouldn't want someone's chest booming like a loudspeaker." Deak clicks his tongue and twists a theoretical screw. "Unless, of course, they don't pay."

This much work would probably suffice for many a sòmnambulist soul in search of fun stuff to muck up. But give a physicist enough time and a free mind and he will go farther—he will start to work on nuclear cold fusion.

Deak is deadly serious about this. In fact, he gave me a research paper on the subject, but it happens to be highly confidential, so you'll have to believe me when I say that some pretty heavy hitters have been working on nuclear cold fusion ever since Drs. Fleischmann and Pons failed all over the front pages in 1989. Deak himself became interested when he realized that certain principles at work in another invention of his—a device to monitor smoothness in machine tooling—might be applied to fusion in solids. In the past few months, he has banded with the likes of Arthur C. Clarke and

Dr. George Cjivanovic (a blood relation to Nikola Tesla and colleague of the famous nuclear physicist Dr. Gamow) to make headway on the world's first acoustically driven cold fusion reactor.

"Things are beginning to look right," says Deak, impossibly confident as he leans back in his swivel chair. "It's starting to sound like a duck. And if it looks like one and walks like one . . ."

Having dropped that bomb, he's almost done with me, but not yet, because somewhere between driving his neighbors stark raving deaf and seeking to provide the world with enough energy for millions of years, Deak makes art. Just now he is sprucing up some canvasses for a show at Z Gallery. And as you might guess, his visual works employ a special technology. He calls his latest technique spectral multi-plexing, though you or I might describe it as the Winking Woolworth Jesus method seen through 3-D glasses.

For the show, colored lights will fade in and out to reveal the alternating images on each canvas. As for style, I don't usually like to submit art with the printed word, but I can say this: his work, strong and accomplished as it is, does not resemble Styrofoam squiggles. And it's quiet. Beautifully, perfectly quiet.

November 24 – 30, 1993

Tactile Solutions

CONTRARY TO POPULAR BELIEF, some inventors actually like giving their ideas away. To hell with the frayed fingernails and the non-disclosure documents; they've just got to get the word out. Take C. Brodeur, for example.

"The idea I'd like to give away," writes Brodeur, stalwart *New York Press* letter-writer, "is one that would be technologically out of my grasp and it shouldn't sit around any longer . . . It's called the Feelphonics Acoustisuit. It makes music for deaf people. It's so obvious I assume variations on it abound. It's not just for deaf people though, because we all love to feel music as well as hear it . . . The suit would be a jumpsuit (with perhaps footsies and hood helmet) with its interior lined with electrodes of some sort (don't ask me man, I can't even use a word processor) which would be hooked up to the stereo."

Brodeur has other provocative ideas, including an electric-shock lollipop, but for the moment let's concentrate on products that would make it past the FDA. (Okay, that would make it past Ralph Nader.) It just so happens that in my dogged research, which never abates, not even during normal business hours, I've come across an inventor who's very much, shall we say, in touch with ideas like the Acoustisuit. And as you might guess, he lives in California.

"What I'm probably best known for is the sex angle that I did," explains How Wachspress, the mastermind behind the San Francisco–based company Auditac, when I reach him by phone. "That was never a commercial success, because the sex market never took off. These days, you never see a sex gadget that costs more than $50."

The sex gadget in question was the Sonic Stimulator, which Wachspress first unleashed on the public back in the '70s. The device fits Brodeur's description closely, albeit without the sartorial niceties. When plugged into the headphone jack of an ordinary stereo, the Stimulator transmits musical vibrations to a plate about half the size of a postcard. Put the plate on your belly, turn on your love light and you're ready to rock.

If the Sonic Stimulator didn't make Wachspress wealthy overnight, it did suggest a different sort of endeavor to a different sort of people, who thought it might be used to make things fly. "I got lost in another invention," Wachspress says, "called the Free Flying Magnetic Levitator. It's the first levitator to fly without a guideway, unlike those levitating trains. Unfortunately, I was involved with the Defense Department, which has made life miserable for me."

An erotic-toy inventor in cahoots with the Defense Department? "Oh yeah," answers Wachspress. "The military is really interested in the tactile."

To hear him explain it, this is not really so far-fetched. After all, the government hopes that someday robots will be able to recognize objects by touch, and anything remotely resembling whatever is bouncing around in their heads will deserve their attention. No, the far-fetched part was that Wachspress thought anything would come of it. The military, perhaps demonstrating its own difficulty in recognizing objects, ended up stringing him along for 10 years, in true love-ya-babe style.

Still, all was not lost. The detour through Hollywood-inside-the-Beltway, while wasting a decade of the inventor's time, inadvertently opened new frontiers for Auditac. For starters, Wachspress allegedly discovered and demonstrated a new principle of physics called longitudinal magneto-mechanical force, and today his company boasts both a Touch Systems Division and a Magnetic Levitator Division. More important, Wachspress' work on magnetic levitation gave him the extra know-how he needed to build a whole new generation of tactile devices.

Making good on his claim that "between sex and not touching, there is a whole world of situations," Wachspress reels off plans for his Touch Systems family of products the way some people reel off the evening's TV schedule. In the next few months, the Sonic Stimulator will be attempting a comeback as an entertainment item, first in mono and soon thereafter in stereo. Once the Stimulator is selling briskly, Wachspress hopes to unveil the first tactile *video* product called the Touch Screen.

"Imagine if you had a screen that had a cloth on it," he muses. "On the other side of the cloth, there's a machine. Then, imagine if someone else had a screen with a cloth on it and you hooked up electronically. You could be feeling each other through the screen." This technology, of course, suggests the need for a recording mechanism, and indeed Wachspress has one in the works. When passed over a surface—say, your left knee—Auditac's tactile camera will record information that can in turn be played back on a Touch Screen.

Even farther down the line, Wachspress plans to make tactile versions of—get this—regular theatrical videos. "What is ultimately going to happen," he predicts, "is that you are going to be able to stick your hand in the screen and feel a movie." I don't know about you, but personally I'm holding out for the tactile version of *Schindler's List.*

That touching is different from seeing or hearing is fairly obvious, but Wachspress has to think about these things more than the rest of us. As he observes, the skin is an intelligent co-ordinate system—it knows which finger is which when you grab a coffee cup—and this mobility must be factored in. What's more, the skin is the only communication system that can both receive and transmit on a common surface. Next time you engage in some mutual consent, try to figure out who is playing the active role. If you were performing an art piece, who would be the performer and who the audience?

With questions like that rumbling around, copyright law could get even messier than it already is. But Wachspress isn't losing any sleep about that. He's more concerned with overcoming the sex angle he helped create. To this end, he has abandoned the idea of selling through stores and instead is busy training his own sales force—sort of an Avon team for the '90s. From the sound of it, he'll need all the help he can get. "People are really out of touch with their own bodies," he says.

So, yes, as you can see, variations on the Acoustisuit do abound. And while there are no plans for a suit per se, there would seem to be a logical opening for one. Virtual reality, after all, relies heavily on the use of a glove. Assuming that people will want to feel virtual objects with their virtual hands (especially after watching their robot slaves play touchy-feely courtesy of the Defense Department), the day when Wachspress' patents are licensed by the cyberheads can't be so very far away.

As for that electric lollipop, Brodeur will be pleased to know that electrotherapy was all the rage in the 19th century, especially among women inventors. Galvanic abdominal trusses, voltaic insoles, electric belts, electrical head clamps for relieving pain—it seems they just couldn't get enough of that juice.

The closest thing to an electric lollipop that I could find, however, suggests the mind of a famous male scientist at work. In my lateral investigations of late, I've come across an item called the Faraday-hand, cited in *Animal Hypnosis*, by Dr. Ferenc Andras Volgyesi. The author, it turns out, studied under the esteemed Pavlov and had something of a knack for hypnotizing lobsters (great pictures here).

But I digress. Shaped, in fact, like a metal lollipop, a Faraday-hand is passed over the skin of the person undergoing hypnosis, causing a mild electric shock. Volgyesi used the device as recently as the 1960s, though the connection to 19th-century electrical pioneer Michael Faraday remains obscure. Until more is learned, readers should refrain from French-kissing wall sockets and stick instead with something tried and true, like a good galvanic abdominal truss.

July 6–12, 1994

Electric Cigarettes

TO SAY THE TOBACCO companies aren't feeling so great is putting it mildly. In fact, they seem more interested in proving Hunter S. Thompson right—as he said, when the going gets tough, the tough get weird. I myself had the chance to verify this maxim when I spoke to a flack at Philip Morris recently.

Me: Hi—yes, I'm curious about your patent for Electric Smoking, #5,249,586, granted on October 5, 1993.

Flack: Yes, well, I don't know what it is, and frankly, I don't need to know.

Me: Mm . . . tell me, what exactly is the function of Public Relations?

Flack: Sir, we deal with a variety of issues every day and if you think yours is the only one, you're very much mistaken.

Me: Ah, I was just curious.

In other words, be a journalist, boy. Make something up. Well, you know, I wish I could, but it's my mission to dog these apparitions to their origins, even if it means blazing a trail through politically sketchy terrain—and on to nowhere at all. So while I waited for the entire patent on Electric Smoking to be published, I went looking for clues. I chased down obscure leads. I pumped distant cousins for information. I actually read a few things.

Among the notable digressions, I called up Dick Sperry, an inventor who comes from a veritable dynasty of patentholders. (His grandfather Elmer, for example, made quite a name for himself fitting out airplane cockpits with gyroscopic trimmings.) I had heard that Dick had worked on a "safe cigarette" back in the '50s, so I thought he might have an inkling why anyone would want to plug in to their habit.

He didn't, but he did describe his safe smoking venture. At the behest of General Foods, he helped develop a cigarette with a clay core. "It had tobacco around the outside," Dick said, "and an inner tube that was coated with an innocuous smoke-producing chemical. You would light up the outside, which would heat up the tube. The chemical in the tube was known to be fairly safe—ammonium chloride, the stuff they use in electric trains. And the tobacco would just burn off."

The coast was clear and the product ready to go when, in a classic case of bad timing, the now-famous warning came out. General Foods did some math, figured out that the Surgeon General had five stars to their four and promptly dropped the project.

"Who knows?" Dick told me. "It probably would have killed people, too. But it was a great project."

I also happened across a reference to some genetic researchers at UC-San Diego, who allegedly crossed a firefly with a tobacco plant to spawn a plant that glowed in the dark. Unfortunately, when I tried to scare up the facts at the New York Public Library, I was told that the magazine cited to me as a source didn't even exist. Further probing revealed only one patent related to both tobacco and phosphorescence: a distant murmur about combustibility.

By now, I had begun to see a vast conspiracy at work, churning out disinformation and thwarting plans everywhere. It took a meeting with Mel Chin, prolific artist and catalyst-at-large, to realize the problem: I was taking the wrong side. The forces of safety are in league with the liars. To get a straight answer, you must court danger.

"I had been thinking about doing a piece about tobacco," said Chin when I dragged him away from his beehive of a studio for a midnight cup of coffee. While traveling from Germany to Corsica recently, he noticed the preponderance of pipe shops in European towns. Soon, a motley array of references converged in his mind: the peace pipe, the cozy professional image of pipes, a spate of bombs blowing up the tobacco business, an exploding pipe in a German cartoon. The result? A "pipe" bomb, which Chin calls the Elementary Object, shaped much like a conventional smoking pipe but packed with triple F superfine blasting powder and crowned with a one-inch fuse. To prevent the casual smoker from blowing his face off, Chin keeps the pipe in a concrete case, lovingly padded with wood shavings.

"It would only hurt the person who lights it," Chin said with some nonchalance. Chin plans to have the Elementary Object shown in various places, including the Havana Biennale. Wherever he ends up displaying it, he doubtless will have achieved one of his stated goals: to avoid becoming "a left-wing spokesman for Philip Morris."

At this point, I think I may have spared myself that fate as well. I did finally get those details on Electric Smoking, though. It turns out the device is meant to heat tobacco without burning it. A cigarette is encased in a metal cylinder studded with tiny internal heaters. When the heaters warm the tobacco, an aerosol is released, thus creating a "fake smoke" vapor (harmless, supposedly) and delivering the tobacco flavor into your mouth.

Clearly, secondhand smoke is a concern addressed by this invention. But the patent also states that the electric, reusable cartridge need not take a cylindrical form, which suggests the potential for camouflage. After all, an Electric Cigarette could take any shape at all. Think about it. Wouldn't that explain the increasing ranks of supposed ex-smokers? I mean, come on. You don't really think those people are walking around with cellular phones to their heads, do you?

And then there is one final clue. Patents often list publications salient to the invention in question, and Electric Smoking makes mention of "NASA Tech Briefs," July/August 1988, p. 31. That's NASA, gentle reader. Does this reference not allude to the problem, bitterly lamented by astronauts the world over, of not being able to open the window when the air gets a little close? Could it be that Philip Morris has overcome the problem of smoking in outer space?

Or is it just the long arm of Hunter Thompson, playing hob with my head again?

November 17–23, 1993

Contraception by Any Means

WE AMERICANS have long been known to deliver the most intimate details of our lives at the drop of a hat. Trap us in an airport during a snowstorm and within minutes we'll be haranguing the nearest traveler with tales of a recent venereal disease, a long ago tryst or—sturdy old chestnut—the eternal lust for lust itself. So it's a bit of a surprise to come across a Yank who, having invented an extremely unusual form of contraception, becomes bashful to the point of non-disclosure when asked to describe it.

"I had a woman reporter in here the other day," says Bill McClenahan, talking to me by phone from his home in Alameda, California. "I realized then that I had never really talked about this thing before. It was pretty embarrassing. I ended up looking at the table a lot."

Fair enough, I suppose, considering. In polite conversation, the device in question could be discreetly fobbed off as Patent #5,603,335. Take it outside and you've got yourself an "intraurethral contraceptive device." But in the company of familiars, you'd have to go with the description from McClenahan's press release, which calls it "a new twist to an old notion—a contraceptive for men that goes on the *inside*."

"My contention," says McClenahan, "is that there's a 'nozzle' at the end of your . . . pee-pee . . . and an enlarged chamber behind it. Well, it just so happens that a coin-like object will slip right in when you turn it sideways, but when you rotate back again, it will increase in its surface area up to 36 percent."

Capitalizing on this convenience of physical properties, McClenahan has concocted a washer-shaped ring that connects to a small tube (lashed to the engorged shaft with an adhesive strip), which in turn empties to a collection bag. Insert the ring into the urethra, give a slight tug to turn it sideways and you're in business. "It stays with *you*," says McClenahan, "and not the girl. If necessary, it could have a kind of tail for extra stability. Anyway, my theory is that as long as it stays on, the sperm will go into the bag. It also seals the man's urethra, so it could protect against some venereal diseases. And it allows for greater intimate contact than a conventional condom does, which suggests that it could have a crossover market."

For the moment, the only existing version of McClenahan's invention is a clay model, built some seven to 10 times larger than scale—a veritable raincoat for a king. What remains to be seen, then, is whether the invention will work for mere mortals. I mean, for one thing, won't it hurt like a son of a bitch?

I think the answer has to be: damn right it will, and as a good American I feel duty-bound to tell you why. When I was an overheated youth, you see, I contracted a couple of low-rent venereal diseases. Chlamydia was one; myco-something or other (the linguistic derivation had something to do with mushrooms) was the other. At any rate, the diagnosis required the insertion of a wire directly into my urethra. This violation didn't last more than, say, a minute or two, but during that time, the doctor had to maneuver the wire so as to scrape a small sample from the walls.

Needless to say, that experience will keep me from ordering a bag of #5,603,335s anytime soon. But let's assume that the user's nerve endings have been surgically removed, or—always a possibility—that he may actually *want* to suffer (more than he already does, that is) during coitus. Even then, there could be problems. For one thing, the device might not remain intact when it's needed most. McClenahan himself contends that "it's the 'nozzle' that allows it to squirt." This suggests that the force of ejaculation would either blow the disc out, or else force it hard up at the mouth of the opening, a state of affairs that's hard to imagine as pleasant.

Unfortunately, I was unable to venture too deeply into such vagaries, because no sooner had McClenahan said the word "squirt" than he performed the telephonic equivalent of a blush. "I'm looking at the table now," he stammered. "Ask me another question, please, David."

Never mind the traditional volubility of Americans. When an inventor balks at the prospect of promoting himself, it's an unusual day indeed. But there it was. Suddenly bereft of our subject, we found ourselves talking about his tour of duty as a class E-4 airman in Thailand during the Vietnam War, his long career at Sperry (including a stint working with the old UNIVAC computers), the beautiful weather in Alameda, even the car horns he heard in the background on my end. Rarely has small talk been conducted in such deadly earnest. Some serious pumping eventually revealed that he had informed his employer of his idea in 1995, so as to clear any proprietary hurdles, and that his boss had laughed him away, saying, "Go make a million if you can." Only the most innocent question about his family life, tendered late in the interview, got him to blurt out that he himself routinely used contraceptives.

Nevertheless, McClenahan did give me something to go on: like the father who freezes up when telling his son of the birds and the bees, he directed me to the cold, hard facts on the inventions that had served as his prior art. And to my surprise, when I went to look them up, I discovered that his invention, far from being proof of the world reaching new heights of absurdity, actually *improved* on these previous contraptions.

In 1973, for example, one Alfred Boyden of San Diego, California, devised what he called an Applicator for an Internal Prophylactic Appliance. I studied the diagrams in this patent for a good half hour without being able to solve the mysteries by which it functioned.

As best as I could figure, the gizmo involved a tube, about as long as the wire that I had once suffered, with a collection bag wrapped around and within it. In the end, I decided it was probably intended for purposes that defy the Geneva Convention.

For S&M pleasures, though, nothing beats Milton J. Cohen's Male Contraceptive Device. Patented in 1980, this invention also makes use of a tube, and like McClenahan's offering, it has a bag for collecting the sperm. This bag, however, is located directly in front of the urethra, as if specifically designed to make a cameo entrance in the vagina. As for the tube itself, well . . . you've seen *Aliens*? You know how the creatures' jaws extend and then extend again? That's the basic idea here, only in reverse: when you push the tube in, it engages on its own threads, collapsing into a cylinder of heftier girth inside.

Lord a-mighty, are we never going to see RU-486 get government clearance? Or does that invention not make for sufficiently juicy airport repartee?

February 26 – March 4, 1997

6

MACHINES FOR THOUGHT

Time Unsprung

IF, AS THE BAND strikes up "Auld Lang Syne," you start to feel like New Year's Eve is a little meaningless, maybe downright arbitrary, you're not alone. Len Toft is right there with you. In fact, the conventions of time bugged him so much he created his own calendar. So it may be occasion for ringing in the new for you and me, but for him it's just plain old 10 o'clock, Sunday, the 9th of Capricorn.

"The Christians were in charge basically," Toft says, explaining the genesis of the conventional Gregorian calendar as we stand stockingfooted in the New York City ashram where he works sometimes. "When they celebrated Christmas, they had a week's vacation, so the new year came a week after Christmas. But my calendar begins on the first day of spring. It's based on nature."

And then there is his pride and joy, the Astrological Clock (Patent #4,337,530), approved by the American Federation of Astrologers and awaiting a backer with pluck and vision. There it hangs on the wall, a deconstructivist's dream. The hands run backward through the course of a 12-hour day (no rascally AMs or PMs). Each hour breaks down into 100 minutes, and each minute into 100 seconds. Toft calls it Metric Time.

Bolstering madness with method, Toft has constructed his timepiece to match the heavens. If the Astrological Clock hangs on a north wall, the hour hand will point east at dawn and follow the sun through the sky until it points west at sunset. (As a bonus, the hour hand also points to the sun's location below the earth at night.) It's a solar compass, if you like. And, just as his new year marks the coming of spring, his first hour—one o'clock Metric Time—marks the coming of daylight.

"The clock we use now is based on a misconception," says Toft. "Mine is based in reality. I call it solid logic. In fact, it may be the most logical clock ever built."

Toft has a preacher's zeal for his invention and a proselytizing urgency in passing the word along. He's even gone so far as to write to Greenwich, England, appealing to the powers that be to adopt his system. "They look for excuses to turn me down," he says. "You should see the denial I get. It's amazing."

Gazing at the Astrological Clock, I imagine an Englishman, very much to the manor born, whose lot it is to defend the names of the hours. One by one, the Tofts of the world approach him, begging him to reconsider. Can't he see? The days are askew; the seconds betray us. And yet the Keeper of Time is bound by duty. To relent is out of the question: "I'm afraid you've got *one minute* to explain yourself, old chap . . ."

Certainly, alternative chronometries are there for the making, since time, like money, essentially means what we want it to mean. All things being equal, we could count by the National Mean Heartbeat ("a pacemaker in every chest") or by a television advertising schedule, with spooky portions of non-time between commercial breaks.

But not so fast. The temptation to measure time by larger, more emotional events remains strong. The Display of the Changing Moon on Watch Face, for example, invented last September by William A. Galison (Patent #5,245,590), counts hours and minutes the conventional way, while an electronic pattern on the watch face mimics the phases of the moon. Now, honestly, if this watch depicted the patterns of your lava lamp, how deeply would it resonate?

Assuming that time is best measured against soul-stirring phenomena, choosing a system may be a question of how you want to feel about yourself. Both Toft and Galison, in aligning the physical clock with an immense periodic motion, have adhered to the principles of harmony. Their clocks have closure.

In this they depart from the current zeitgeist, however. Nowadays, we are not so inclined to believe in such niceties as a cyclical universe. We want time to be honest and a little bit brutal. Thus, the appeal of the National Debt Clock, with its Keynesian departure from reassuring cycles, its incessant slide into eternity.

Seymour Durst, inventor of the National Debt Clock, may be eternal, but he is far from incessant. When I spoke with him on the phone a few weeks ago, he sounded about as relaxed as a person can be. He talked in a free ramble about his timepiece, which hangs on a wall at Forty-third Street and Sixth Avenue, fiercely broadcasting the federal debt.

"The numbers increase at $13,000 a second," he told me, "so the computer moves it up at that rate. But the debt doesn't move constantly. When it gets pretty far out of order, we adjust it."

I was curious to know how often an adjustment had to be made. "It's strange," he answered, "in the month of October it only increased by 10 or 11 billion, and then the next three weeks it increased, I think it was 65 billion. That was a very major increase."

In other words, Debt Time is plastic. It can speed up, or slow down. And it's dynamic: it asks for a response from a citizenry hell-bent on its own destruction.

Meanwhile, David Kendrick of Berkshire, New York, has come up with a measurement even more sobering than your standard forecast of economic squalor. Kendrick's patented Reverse Watch measures not what time it is, you see, but how much time you have *left.* Using actuarial tables, you can determine your life expectancy down to the second, and act accordingly.

Kendrick himself bears out the fast unfolding maxim that a foreboding of doom will keep you calm. When I spoke to him by phone,

I did not get the impression he was frantically monitoring the seconds. In fact, he seemed even more relaxed than Durst. Ironically, he had to cancel our scheduled interview after a friend of his died unexpectedly. (Whether this friend wore one of his watches, I suppose I will never know.)

Each of these chronometries has its merits, and each would make a fine international standard, I'm sure. Still, I think I know one that subsumes the lot of them. Patented by forces I wouldn't care to petition, the Mayan calendar uses both sun and moon for its bearings and even takes social behavior into account. What's more, in the Thompson correlation (named after Sir Eric Thompson, a Mayanist and, one assumes, another Keeper of Time), the Mayan calendar predicts that our recalcitrant ways will bring about the end of the world on December 23, 2012 A.D.

A quick calculation says that gives us a whisker less than 19 years to do whatever it is we're going to do, no matter what the actuaries may claim or where your hour hand may point. My advice? Think about the Mayan calendar. A lot. It's bound to make you more relaxed than you ever thought possible.

December 29 – January 4, 1994

Alarms and Locks for All Things

FOR THE PAST FEW MONTHS, I've been pushing the big themes—the fate of the universe, the human ambitions that clutter it up, etc., etc.—and neglecting the workaday gadgets almost entirely. Don't ask me why. I know for a fact that a paper clip touches the American soul in a way that leaves anti-gravity machines spinning into irrelevance. Such is the glory of our diminished expectations, and by extension (yes, I admit it), the onus on me to report them. And so, in the great tradition of confession and penance, I've gone back to the patent archives in search of the little things that thrill.

Funny thing is, the small inventions sometimes reveal bigger trends in spite of themselves. Or rather, they reveal multiple trends, all swimming about in the kind of stew that gets chaos theorists salivating. In any given week, no single pattern asserts itself among the issued patents with absolute certainty; yet discernible patterns are always there to be found. Today, one of these patterns reminds me of a speech made by George Cortelyou in 1933, on the occasion of inventor Elihu Thomson's 80th birthday.

"[B]ehind the marvels of science and invention," toasted Cortelyou, "lies an exceedingly complicated and artificial system . . . When all its parts are working smoothly we are not conscious of the intricate mechanism by which it is governed, and accept its benefits as a matter of course. But when something happens to throw it out of balance—it may be one of a thousand things—causing a slowdown or stoppage of any one part, the whole machine functions poorly or not at all."

Now, I can't vouchsafe Cortelyou's opinion when it comes to writing (an imbalancing act if there ever was one), but I can surely imagine times when our "complicated and artificial system" is vulnerable to havoc, and thus in need of a patented fail-safe device. Nor, from the looks of the files, am I alone on this matter.

Take the plight of the air controllers for example. These sturdy souls live a lifetime or two every time they jockey up to a radar screen. Jets take off and land every which way, stack themselves severally and, of course, move at blistering speed. The controller, meantime, must track the whole circus on equipment installed when Nehru jackets were hip. If he looks away for a second, the blips on the screen turn to gibberish.

In recent labor battles, management has shown some credible evidence that controllers actually do a better job when their equipment is *not* heavily automated. Seems human alertness has its virtues even outside the watching of TV commercials (a point that the Metropolitan Transit Authority, which is planning to automate parts of its subway cars over the next few years, would do well to heed). Of course, brand-new *non-automated* equipment for controllers might not be a bad idea, either. But while we're holding our breath for this, Harris R. Lieberman's Vigilance Monitor System, patented this month, poses a reasonable alternative.

The Vigilance Monitor System isn't a Big Brother device so much as a sophisticated "Corpse-Retriever"—the alarm designed by one of Thomas Edison's muckers to keep himself awake during a particularly intense research project. Lieberman's patent claims are broad enough to cover almost any arrangement, even those already a century old, but we can see the big picture at least. The controller receives a stimulus—I don't know, maybe a big paddle whacks him on the forehead—and then special sensors record his reaction time. Too slow a reaction and he's yanked off the line for a spell.

Needless to say, in 20 years, the Vigilance Monitor (to say nothing of the monitor to monitor the monitor) will fail just as often as every other piece of aviation equipment does. And when that happens, you'll want to dig out your variation on Patent #5,683,130, otherwise known as the Underwater Vehicle Recovery Method, invented by Raymond N. Pheifer of Lerna, Illinois.

The Underwater Vehicle Recovery Method is one of those "doh!" ideas: so simple you can hardly believe no one thought of it earlier. In fact, it's nothing but a gigantic slip knot. Take your loop, what sailors call the bight, throw it around any one of the wheels—and pull. As Pheifer sees it, the Underwater Vehicle Recovery Method can be used by trained rescue workers after minimal training "under conditions of zero visibility." He also points out that the method "requires minimal capital investment"—always a plus, as I say, if you happen to live in any year after 1989.

Well, I'm having my fun at the expense of decent folk, I know. If cornered by a humorless personage—say, a psychiatrist—I'd have to admit that both of these inventions strike me as worthy endeavors, because they address slowdowns or stoppages that might otherwise cause death in sizable numbers. In situations calling for heroism, precautions amount to saving graces. Situations calling for laziness, however, are another matter. Apparently, we Americans—and thus our inventors—will not rest until every last object of our affections has been battened or bolted down.

I can think of no better illustration of what I mean here than the Adjustable Toilet-Lid Closing and Lock-Down Device. When I first read this title, I thought it referred to the toilet *seat* and registered a tiny thrill. Finally, I thought. Someone has settled the war of the sexes—not by artful compromise but by locking the sucker down. Silly me. In point of fact, Gene Doyle Burt of Sandy, Utah, has invented a device for locking the lid, said lock requiring "adult supervision to use, thus providing safety and general hygiene for infants, young children and household pets."

Try as I might to put on the glad face, I find this image disturbing. Sure, security is one of the fastest-growing sectors of the economy, but have we truly reached the point where a potent symbol of transgression must be made more so? Has little Joey really been drinking from the can all that much?

Such a thin line between prudence and distrust, and so easily breached. Indeed, barring a clever sensor to detect where that line lies, every spare moment of existence may soon be attended by a signal to remind us what we are seeing or doing. If I were of a paranoid persuasion, I might even suspect some dastardly crew of trying to loose Hollywood scriptwriting techniques on the world. ("So . . . that means we're sitting on a bomb?" "That's right Hutch. And viewers can see it right there on the lower half of their television screens.")

Crazy? Maybe. But peruse the patents of late '97 and you will see: the endgame is already in sight.

Say you head out for the creek with pail and rod, intent on a day of dappled reverie. Casually, you kick off your shoes, tie the line to your toe and pull your hat over your eyes. As the afternoon slips by in that beautiful way that afternoons do, will you begin to feel that you've gotten away from it all—the sirens, the intricate mechanisms, the toilet-seat locks? No sir, not by a fair bit. Because when the trout finally bites, a signature tone will ring out, verifying the working presence of Andrew Corbiere's Patent #5,682,703 . . . for a Fishing Rod Alarm Apparatus.

Reader, promise me this: you will not pack a cell phone on your next fishing trip. Promise me that you will not bring a geostationary satellite for watching the top of your daughter's head, or a blue-sky simulator or a pair of sunglasses that tell you what time it is in Madagascar. And above all, take a solemn oath—just for me, on pain of death—that you will never, ever, knowingly use a Fishing Rod Alarm Apparatus. Unless, of course, you happen to be casting for jets.

November 19–25, 1997

The Universal Board Game

IN THE BEGINNING of Jorge Luis Borges' story *The Aleph,* the character of Carlos Argentino Daneri appears to be the worst sort of dilettante. He is hard at work putting the entire universe into a poem, the brilliance of which he is only too eager to relate at the slightest prompting. By the end of the tale, however, the narrator has put on the fool's cap, and Carlos Argentino has been recognized as a genius by all.

Richard Gosselin could probably find some interesting parallels between the tale of *The Aleph* and his own life journey. Not because he's a latent Argentine writer, but because his board game, *Reality Rules,* attempts to describe the whole of existence.

"This game is the game that Stephen Hawking or anyone in recorded history has been going on about," Gosselin told me when we met recently. "Finding a material order and then realizing that there's something present that's thinking about it."

A slight, sandy-haired man with the look of a weathered sailor, Gosselin was inspired to tame his unwieldy subject when his son began to ask the tough questions about life. "I thought I might convey something of what I know," he explained, "about this apparition of reality."

The production was hardly a snap, however. True, he made the seven boards in a matter of months. But he also found himself frequenting the local car wash, it being the only place he could get enough water pressure to clean his silk screens. And with only a few screens at his disposal, he had to destroy the images for a completed board before he could move on a new one.

A $1500 loan from a friend saw him through to the finish, at which point he copyrighted his magnum opus.

That was 1981. The night before we got together, I had the chance to look at one of the three existing prototypes and to note the careful beauty of the boards—a galaxy here, a chemistry diagram there. I didn't have time to commit much to memory, but I could readily see that *Reality Rules* had an approachable logic. Game One, for example, mimics the table of elements. Pick up an electron and advance a square as you try to gain atomic weight. In the second game, try to achieve life as our molecules take shape. The succeeding boards progress through the food chain, onward to the ever-fascinating species of human, to the stars beyond and, finally, to the most difficult game of all—Game Seven, or Logicon.

"This is the part I'm still working on," Gosselin conceded, "because it's important for me to understand the end." The basic premise of Game Seven, he said, remains unchanged: the winner gets to interpret the meaning of the six previous games. What is new is the "time warp" card, the meaning of which only he, the author of the game, will ever know.

"It's a simple concept," he said. "Anybody would understand it, but I believe it is the key to the understanding of time."

He reasoned it out like this: the express purpose of the game was to help people navigate reality more nimbly. So how would he know if he, Richard Gosselin, was winning? Well, he decided that one proof of his mastery would be the general prospering of the game itself. After all, there was always the chance that the boards could revert to dust in a closet somewhere, leaving their author to perish in obscurity. But as long as he knew the meaning of the secret card, he could keep his bearings in this chaotic old world, no matter what fate his creation might suffer. Mastering the secret card meant winning the game off the board.

In a kind of fringe benefit, he said, the mystery of the "time warp" card might encourage the other players—assuming that someone besides Gosselin gets to play it someday—to ponder the possibility of their own secret card. "When you see someone you think is all right, it's like they know their own card. And maybe that gets you to thinking: 'What's the meaning of their card? What's the meaning of my card? Do I have one?' "

I thought about this for a minute. "Can a person know his card," I asked, "and be a killer?"

Gosselin was clearly struck by this remark. He measured his response carefully. "I won't answer in the negative, but I certainly won't answer in the positive, either. I can tell you, though, that the nature of evil is definitely incorporated into the game."

Throughout the evening, Gosselin waxed eloquent about gambling, Heidegger, chopping wood—all fitting subjects, of course, when discussing the totality of things. He also imparted some ideas about the environments in which *Reality Rules* could be played. As he saw it, children could play easier versions of the games in an educational setting without the weightier topics of alienation, exile and death ever threatening to intrude. Meanwhile, a version of Game Seven could be played in a bar on, say, East Houston Street.

"They wouldn't necessarily understand the deeper meaning of it," Gosselin said, with a raconteur's confidence now. "It would be more like something that would hit them as something to do. They could see it on the wall, and maybe the loser has to buy the winner a drink. But later, it would make them wonder what it was all about. Then it becomes like a map—a map of the mind."

To tell you the truth, I find Gosselin's pursuit appealing, all the more so for its implausibility. Before we talked, I had worried that he might present a load of neurotic demands. I had even wondered if his prototype could be a bomb. But having met him and wormed my way through some of the intangibles of *Reality Rules*, I must admit he's got me thinking. Could I hold my own in a black hole? Would I want to know anyone else's card? On the board, off the board—can a person know his card and be a killer?

February 16 – 22, 1994

Secret Ingredient Revealed

WHENEVER I HAVE the tiniest bit of idle time on my hands, my thoughts inevitably run to the subject of scams, hoaxes, forgeries, con games—all those things that make up the great tradition we call American. And because I've been downing the occasional can of Coca-Cola lately (for reasons that remain obscure even to me), my thoughts have been running in particular to Coke's secret ingredient, the fabled 7X. I've become convinced, you see, that 7X is a complete fabrication. As in null, *nada*. Simply not there.

The cynical calculations of a tired old scribe? Could be. But when a friend laughed and told me that I was fobbing off a "very '90s idea," I could only give my wholehearted assent. A '*90s* idea indeed . . .

A hundred years ago, when Coca-Cola first appeared on the planet, the patent-medicine industry was functioning at a level of near pandemonium. Products with names like Dr. August Koenig's Hamburg Breast Tea and the Genuine Dr. C. McLane's Liver Pills and Vermifuge wold arrive in remote village squares along with troupes of acrobats, jugglers, equestrians and Punch-and-Judy shows—all of which was merely the warmup for the pitchmen themselves. At the signal moment, Calculator Williams would take the stage and fire off complex mathematical calculations at breakneck speed. Big Foot Wallace, meanwhile, preferred to address his crowd at gunpoint. With people like that working the circuit, you could get tired looking for something more exciting to see.

The most successful medicine shows of the 1890s, or the biggest anyway, were those put on by the Kickapoo Indian Medicine Company, brainchild of John E.

Healy and Charles Bigelow. In 1879, Healy was a door-to-door salesman of vanishing cream, shoes and a liniment called King of Pain. (Sting, where is thy death?) After a brief stint as the organizer of Healy's Hibernian Minstrels, who in blackface were hardly Hibernian, Healy hooked up with one Dr. E. H. Flagg, a pitchman who sold Flagg's Instant Relief while playing the fiddle. The two men then hired Bigelow—a self-styled Texan with a sombrero and a quick tongue for Indian lore—and the action commenced.

When the first Kickapoo shows appeared in 1881, they involved a smattering of hired Indians (members of the Sioux nation probably) whose job it was to stand around a kettle and ladle out its contents to any audience members who had shown enough foresight to bring their own bottles. Gradually, this scheme became more sophisticated. Assuming the title of "Indian agents," Healy and Bigelow claimed the sole right to sell authentic Kickapoo product and, while the Indians remained onstage as mere ornaments, openly confessed their own ignorance as to the ingredients of the savages' remedy. Then again, how could it have been otherwise? As Mark Twain mused about stage Indians in general, the Kickapoo was "an extinct tribe that never existed."

As audacious as Healy and Bigelow became, their antics didn't seem to faze the public terribly much, and eventually they were able to expand beyond their wildest dreams. By 1890, they had 31 Kickapoo troupes in Chicago alone and an estimated 800 Indians in their employ. In New Haven, Connecticut, they fitted out the Kickapoo Medicine factory as a dimestore museum and, from their offices there, began putting out a soporific magazine called *Kickapoo Indians: Life and Scenes Among the Indians.*

Ironically, the other big success story of the Gilded Age came with no show attached at all. This was Lydia Pinkham's Vegetable Compound, a hodgepodge described on the label as consisting of "Unicorn root, Life root, Black cohosh, Pleurisy root, and Fenugreek seed macerated and suspended in approximately 19 percent alcohol, for preservative purposes." Pinkham herself, it should be noted, was a member of her local temperance society, so what she was out to preserve—the tonic or the taker—remains ambiguous. Nevertheless, her compound was a hit with the ladies, who took it for their various prodigal Victorian ailments, and it remained on the market all through the Prohibition years, until 1975. (Pinkham defended her alcohol levels by pointing to the number of alcoholics who drank it and became violently ill. Still, that didn't stop railroad workers from breaking into her shipping crates in the '20s to get at the potent brew.)

The various Kickapoos and Pinkhams of the world had already established the limits of liberty when John S. Pemberton, veteran of the Confederate army and journeyman druggist, staked his big claim. In fact, many of Pemberton's curative potions were already famous throughout the South by this time: Extract of Stillinger, Gingerine, Globe of Flower Cough Syrup, Indian Queen Hair Dye, Triplex Liver Pills and Pemberton's French Wine Coca. But lasting fortune eluded him until 1886, when he modified his French Wine Coca (removing the wine and adding caffeine and extract of cola) and patented it as a nerve stimulant called Pemberton's Tonic.

Pemberton's reason for altering his nostrum was ostensibly medicinal—his goal was a cure for headaches—although he may well have taken the wrath of the temperance movement to heart. In any event, the rest was the stuff of legend. He brewed the first batch of his new syrup in a brass kettle sometime on or around May 8, 1886, in the backyard of his Atlanta home, at 107 Marietta Street. That same day, he took a jug of it to Jacob's Pharmacy, where Willis E. Venable's 25-foot counter served as a soda fountain. Venable liked Pemberton's brew so much that he started selling it on the spot. The name (Coca-Cola Elixir and Syrup) and the carbonated version followed soon after.

Coca-Cola retained its medicinal status during its early incarnations. Asa Candler, owner of the company from 1886 until 1919, billed it as "The Wonderful Nerve and Brain Tonic and Remarkable Therapeutic Agent." In 1890, Candler wrote to a doctor in Cartersville, GA: "The medical properties of the Coca plant and the extract of the celebrated African Cola Nut, make it a medical preparation of great value, which the best physicians unhesitatingly endorse and recommend for mental and physical exhaustion, headache, tired feeling, mental depression, etc."

Unfortunately, the deleterious effects of cocaine were pretty hard to ignore, and by 1900, reports of Coca-Cola addiction had begun to surface. (A Dr. Purse of Atlanta, for one, described a teenage messenger boy who arrived at his office jobless and dissipated—the result of a daily habit of 10 to 12 glasses.) This also happened to be the moment when the temperance movement and the law coalesced to pass the Pure Food and Drug Act of 1906. Candler, his finger to the wind, had taken the cocaine out of Coca-Cola not long before that. And surprise, surprise: only then did the company begin to speak of 7X, a secret ingredient known only to a handful of people.

When I'm feeling particularly nasty, I'm apt to regard Coca-Cola as the equivalent of the tobacco industry. Sure, 7X falls under the category of trade secret, which means that it's closely guarded rather than patented. But who else, I carp, is permitted to hide behind the catchall of "natural flavors" these days? (Answer: no one.) When I get to framing the matter this way, I generally start gnashing my teeth and forming arguments about pharmaceutical delivery systems.

Thankfully, this mood usually doesn't last. All I have to do, in fact, is consider that Pinkham grew rich without a show and that Healy and Bigelow did the same without any ingredients, and my funk immediately begins to lift. And to draw that thought out to its logical conclusion—that 7X must be its own kind of Kickapoo secret, sprung on the world with the corpse of Big Foot Wallace not even cold—well, somehow that pleases me more than anything.

October 8 – 14, 1997

Hole in the Head: A Love Story

LATELY, I'VE STARTED a tradition, which I hope will be widely adopted, of keeping a CD index. By CD, I don't mean a place to put your money, nor do the letters stand for compact disc. They stand for "cognitive dissonance," and the index is meant as a general marker for tracking the loss of one's wits.

My latest entries have mostly to do with the news. One local TV station, for example, has taken to boasting of itself as the place "Where the News Comes First." This slogan got me wondering if other news programs were leading with stories on hospital corners. But I shouldn't have been so naive. When I checked around, I discovered that, far from being outrageous, this first station was tame compared to the competitor that broadcast the "News Ahead of Its Time." News ahead of its time? I made a note in my CD index and considered myself on a roll.

And I suppose I was. Thinking that there might be a radio station somewhere delivering news that never happened, I turned on my transistor and heard a woman belting out what sounded like old Bessie Smith tunes. So far, no paydirt. Then, during an in-studio interview, this same woman interrupted an otherwise dull repartee with the confession that she was taking mood stabilizers. A blues singer! On Prozac! Another entry, with a flourish.

Of course, past the literal description of events, an index like this begs the question of quantifiable differences. Indeed, whenever I start muttering that overheated promos and underheated divas are evidence of a world gone insane, I have to remind myself that such things rank somewhere around two or three on a scale of ten. Case in point? Holding down the far end of the spectrum, without a contender in sight, are the things people have elected to do with a simple tool called a trephine.

A TREPHINE IS BASICALLY a crank-operated circular saw that works like a corkscrew. Protruding through a saw-toothed circle is a spike, meant for getting a grip on whatever you intend to cut. Crank the trephine and the circular saw turns. Keep sawing and you'll get a circular disc, which, thanks to the spike, you can pull straight out.

You can buy a trephine at most surgical-supply stores, because they're used for operating on cerebral ulcers and the like. But the mere fact that certain devices exist is a source of temptation, and people will invariably find novel uses for anything that's been invented. For example, there are people who have used this gizmo to drill holes in their own heads.

The story of self-trephination (or self-trepanation, if you're British) traces back to the visionary philosophy of one Bart Huges. A Dutch doctor with the juice of the psychedelic era in his veins, Huges had a revelation in 1962. Consciousness, he said, is related to the flow of the blood to the brain. As the human race evolved and began to stand upright, blood stopped making it all the way up to the head, thus causing certain qualities of awareness to atrophy. Today, said Huges, the only vestige of the unrestricted brain is the opening in babies' skulls, which closes as they mature, leaving them—meaning us—bereft of their most precious creative powers.

The solution, which Huges undertook on himself, was to open a hole in his skull with an electric drill. Pleased with the results, he then began to preach the virtues of self-trephination to others, for which he was promptly whisked off to an insane asylum. And that would have been the end of it, except that before he was relegated to obscurity forever, he made a convert of Joseph Mellen.

Their meeting took place on Ibiza in 1965, during what seems to have been a prolonged acid trip. Mellen immediately warmed to the idea, and upon returning to his native England, made preparations to undergo the cure. He bought a trephine and some needles for administering a local anesthetic. Ultimately, the needles proved to be too fragile and they kept breaking, so the following day, he got a tougher batch. With these—and a tab of LSD for good measure—he was ready to go the distance.

Mellen was successful at making an incision into the bone, but was unable to make any headway with the sawing portion of the program, for the simple fact that he couldn't get any leverage. Frustrated, he called Huges, who agreed to come from Holland to help. Huges, however, was barred from entry in Great Britain as an "undesirable visitor," and the project was stalled . . . until Mellen's friend Amanda Fielding agreed to take his place.

Between them, Mellen and Fielding had no medical experience whatsoever, and they were not exactly entertaining the idea of a stroll in the park. Then again, didn't Louis Pasteur inject "hydrophobic" boys with the rabies virus and receive the laurels of the civilized world for his troubles?

Armed with what must have been some justification on that order, Fielding re-opened the wound Mellen had made. Using all her strength, she then succeeded in getting the spike to take and the saw teeth to bite. Mellen took over the cranking from there—which proved to be a bad idea, because in mid-operation, he suddenly fainted. When he arrived at the hospital, mortified doctors told him he was lucky to be alive.

Mellen went through a few rounds of psychiatry and a couple unrelated distractions before he tried again. This time, he found the old groove easily and began sawing, with Fielding standing by. According to his own testimony in the book *Bore Hole*, "After some time there was an ominous sounding schlurp and the sound of bubbling. I drew the trepan out and the gurgling continued. It sounded like air bubbles running under the skull as they were pressed out. I looked at the trepan and there was a bit of bone in it. At last!"

It took one more attempt in 1970, with an electric drill, for Mellen to be absolutely sure that he had opened a hole large enough to free up the untapped potentials of his brain.

"This time," he wrote, "I was not in any doubt . . . A great gush of blood followed my withdrawal of the drill. In the mirror I could see the blood in the hole rising and falling with the pulsation of the brain."

After that, it was Fielding's turn, and as a measure of the confidence they had gained, they decided to film the occasion. *Heartbeat in the Brain* may not be a snuff movie, but it comes awfully close. Audience members dropped off their seats "one by one like ripe plums" at a London screening of this film, which shows a blood-spattered Fielding, a patch of hair shorn, sawing away to a new-age soundtrack.

Neither Mellen nor Fielding has ever recommended self-trephination for others, though Fielding has vigorously lobbied Parliament to legitimize trephination as an out-patient medical procedure. But the kicker of it all—and the point where their story becomes truly worthy of cognitive dissonance—is that, on one level at least, their experiment worked. Today, they share a flat, which they've peopled with two children. They run an art gallery, called the Pigeonhole Gallery, and they've started their own publishing business. Above all, they're *happy*. No blues, no Prozac—they give new meaning to the phrase, "You can't argue with satisfaction."

Learning about the exploits of Huges et alia did little to make me want to loose the pulses of my own beleaguered cranium. Still, I wonder. What exactly does it mean to get a result like that from a piece of equipment with only two or three moving parts? Open your CD index, and put it down as a ten.

January 29 – February 4, 1997

A Confidential Scheme

NOW THAT WE'VE established that there is a thing called sex in America, I feel compelled to lift my voice in song with those who have discovered a thing called statistics in America. In fact, this prurient and utterly sordid activity has long been prevalent throughout the country (though admittedly, Protestants have been consistently more likely to have statistics than any other demographic). Recent studies, however, show that people are having more statistics than ever—and this can only be attributed to AIDS. Where once sweet fables were woven into the many stages of undress, the exact voice of risk now commands the proceedings, with its endless litany of factors, figures and ridiculous partner histories.

What can be done to ward off this ghastly actuarial frenzy? Curing AIDS would be one way, of course, but that, sad to say, would require numbers and charts and other unwholesome devices. So while we're waiting for this nasty virus to just—I don't know—go away, we might as well listen to Alfred Nassim, a software developer who has traveled the world and seen much.

Nassim recently posted the broad outlines of Patent #5,108,131 on CompuServe as "An Ethical System For Reducing Health Risks During Sex," and that's almost what it is. More accurately, it reduces the risk *before* having sex. (Once you're in for the count, Nassim has little to say about what you do.) It works like this. After receiving a card with a photo I.D. and a card number, the user—or the used, to be fair—tiptoes off to an approved testing center, where his card number goes into a computer network. Our hero is then tested for any one of the many sexually transmitted diseases, to return some time later for counseling and results. From then on, he can call a special phone number, punch in his card number and an associated secret code—a PIN—and receive the results in number form.

At first glance, the system seems to change matters little, except maybe to make confidentiality a little less stringent. But in fact, there are hidden virtues. For example, most people ask the important question only after they're three sheets to the bedpost. If two prospective partners agreed to exchange card numbers and PINs, however, they could dispense with their niggling doubts before they arrive at the point of no return.

Partners could also agree to have their card numbers linked in a sort of joint account. Then, if one partner became infected by a third person (either during the relationship or after), the possibility of infection would show up, sans names, on the other partner's "account." This feature would eliminate the need to trace partners, surely the job that gets the lion's share of equivocation in the current system.

Most bright young ideas never see the light of day simply because they cost too much in the start-up phase. Nassim's system is unusual in that the technology is already in place, making it cheaper to implement than, say, your average coffee-bar chain or convict-roast. This fact did little to send any quick cash Nassim's way, however. Moments after he posted his idea, the CompuServe thread went berserk. One respondent went so far as to equate it with Nazi Germany's practice of tattooing Jews. "Pink is a nice color," he smirked, "what about putting a pink triangle on those who did not buy our 'sex' card?"

If the others just blinked at that, they still leaped into the fray without hesitation. Indeed, this is the kind of idea that, even without the AIDS angle, tends to get loophole enthusiasts reaching for their lassoes. "What would my wife think if she saw a . . . card in my wallet?" asked one man, who presumably had something going on the side. Another suggested that the ownership of such a card would telegraph guilt, not realizing that, unlike a credit or ATM card, Nassim's plastic is to be presented only at testing centers.

A more astute participant ventured that the card would be valid only on the day of the test. Nassim rebutted by saying that the risk of HIV transmission is reportedly between one to 500 and one to 1000 sex acts, making the risk between yearly checkups rather low. I think you can spot the fallacy here, though: if the risk between yearly checkups is low, then why even bother to get the second test? The same reasoning would apply with every passing year. Measure the risk in annual chunks and there's no need to be tested at all. (There, you see? It *did* just go away.)

Nassim's system doesn't iron out every wrinkle in the current system, either. Those unwilling to take an HIV test now would probably not become any bolder once computers were in the mix—especially at $20 a card—and dishonest PWAs (statistically, anyone who hasn't been tested) could still cover up their status by stealing a card and learning the PIN at their leisure.

These are fringe considerations, and probably wouldn't alter the situation in one direction or another. But in one respect, a card could make matters much worse. While doctors would probably be loath to violate the ethics of confidentiality, whoever was in the business of handing out the cards would remain an unknown quantity. Clearly, the only way to do it would be to sell the cards at lottery-ticket outlets, no questions asked.

Nassim is a smart guy, and like many smart guys, he hopes his rationale will stem an irrational tide. Who knows? Maybe it will. At worst, it seems like an improvement over the present state of affairs (sorry), if not for this generation, then at least for the computer-friendly one on the way. Nassim says as much, and there's a compelling corollary behind such a hope. "Fortunately," he writes, "we normally start our sexual careers with a clean sheet. If we continued having sex with others who also have a clean bill of health, we would not get infected."

Simpleheaded as that may sound, it's a line of reasoning worth exploring. What would happen, for example, if every U.S. citizen were to be given a card and an anonymous HIV test *at birth*? With the testing at a full 100 percent no one could even suggest that any civil rights were being wronged, because the presumption would be that every child would test negative. In the early years, tests would be administered only after extreme events such as blood transfusions. Then, when school entered the picture, the testing could be stepped up to something like once a year. This would be to appease the one or two paranoid parents on the roster, but it would also set the stage for the more worrisome teenage years.

Of course, children who tested positive would be under no obligation to disclose their condition to others any more than they are now—or even to know about it, if their parents so chose. And as the precious days flitted by, tomorrow's burgeoning adults would be able to conduct their puberty with the awkwardness that is their birthright, until such time that mom and dad chose to sit their little darlings down and have a heart-to-heart on that most daunting of subjects, the one known as *probability*.

November 30 – December 6, 1994

Tensegritous Thinking

IT'S PROBABLY a measure of our society that, as my wife Leslie has pointed out, there is no adjective for the word *integrity. Integral* clearly doesn't make it. My thesaurus indirectly offers *upright, honest, virtuous* and a few others, but none of these convey the sense of economy that integrity does, so for the moment, I suppose I'll have to make do with Leslie's second line of defense and say that Brian Higgins is an *integritous* thinker.

For example, as we sit in the clear sun of Seventy-second Street, at a wrought-iron table outside an overwrought cafe, he's building a case that copyright as an idea may be valid because in certain circumstances it can be invalid. He slaloms around a couple more moguls on the slippery slopes of logic, his voice crisp and void of "likes" and "you knows," his eyes curiously bright for being so dark beneath his silver hair. His smile, which flashes out broadly from a surprisingly compact mouth, possesses some remarkably clear energy. Certainly, nothing about Higgins fits the profile of a man who moves from town to town every few years, putting up in cheap hotels. Nothing, that is, except his explanation for the journeyman's life. "I do things in a consistent way," he says, "and I get consistent results. The longer I stay in one place and keep doing things in the same way—I keep beckoning the same results from my circumstances. So it becomes like a spiral and finally I have to break out. I have to change my circumstances completely. The great thing about New York, though, is that it's so huge. I give it about 10 years. After that, the world is the next large set of circumstances, and that's good for a lifetime."

But before we get to the world, Higgins has his tale to tell about copyrights. It started back around 1981, when he discovered Buckminster Fuller's principle of tensegrity and fell head over heels for it. "For years, I wasn't able to sleep at night," he says as he gives me a large sketchbook to peruse.

Now, if you were to run to your Fuller file and get a definition of tensegrity from that wily old inventor, you'd end up thanking the Microsoft guidebook writers for their clarity, so let's try to be plain. Tensegrity is an engineering principle. Its basic elements are rods and connecting strings. The strings connect to the rods in such a way that the rods are suspended by the strings, no matter the angle. Typically, none of the rods touch one another, though they can. Make the strings elastic and you've got the Tensegritoy, covered in these pages some months back. Take a Tensegritoy in your hands and push one rod. All the other rods move. It's an honest-to-God interconnected unit.

"These drawings are great," I offer, looking up from his sketchbook glad that I don't have to lie. There are tensegrity designs for a human body, sculptures, a host of objects I can only characterize as stylized snowflakes. Each one is drawn in black pen; many are washed with gray watercolors. The whole book feels canonic, complete, which may explain why Higgins describes its contents as the plans for a life's worth of work.

As it turns out, first on the list for the rest of his life—illustrated in the beginning of his sketchbook—was his idea to apply tensegrity to planes rather than rods. This inspiration led to a display sign made of several surfaces stacked one on top of the other, with each surface at a right angle to the one below it. The kicker was, none of the surfaces touched one another. They were supported—well, let's give it the big one—*tensegritously.*

"Here's a slide of it," he says, handing me a plastic-jacketful of transparencies. The display sign looks like a work of art, which it should, since Higgins is a practicing artist who has had shows as far afield as Tempe, Arizona. But it's also a good example of function meeting form: whatever is written on it—"Buy Yarn," "Come Hither," whatever—will be visible from every angle.

And there's the rub on copyrights, because when Higgins gussied up one of these display signs with printed matter on it for the Register of Copyrights, he obtained a copyright, no problem. But the printed matter belonged to someone else, so Higgins went a step farther and tried to copyright the display sign sans alphabetic blemishes. This version, alas, was turned down. To Higgins, this suggests that only the writing qualified for legal protection. But who can know for sure?

"Here's what you should try," he tells me. "Copyright one of your articles just as it's published. Then submit the same article under a different name and see what happens. Keep changing it until it's accepted. That would be a kind of proof of what's original and what isn't—what's false and what's true."

Of course, Higgins could try to do this himself, but lately he's been too deeply immersed in the many other possibilities that tensegrity suggests. He's been furtively guarding a brown box ever since we sat down; now he opens it and produces his tensegrity model. It's a truncated tetrahedron, or what we lay people call a complex but totally cool shape. The drama of the unveiling is dampened somewhat by my having seen one before, but it needn't be, since Higgins' ideas about what it can do are all his own.

"I've been thinking of using this model as a teaching tool," he explains, still exercising the intense organization of his smile. "At one point I was reading [John Maynard] Keynes, and a lot of it went right by me, but I liked his idea of the elastic economy. Now, this model demonstrates elasticity. When most people think of structure they think of girders, of rigid structures. This is a flexible structure. So I'd like to be able to explain economics with this model."

He balances the model on all five fingers, a crystal ball of angles. "This is the global economy," he declares, returning to his largest set of circumstances. "It's all we've got, no more, no less. When something happens in one part of the world economy, every other part is affected. So I'd like to be able to use this model, come up with some values for it and put it out on a computer spreadsheet."

In the beginning of our conversation, Higgins mentioned that he thought of what he does as more improvisation than invention, and I take that as my cue to jump in here. Together, we assay the analogies. Maybe the strings are the economic exchanges, and the rods are the various business units—stores, households, corporations.

But apparently, this is all Higgins needs to begin spooling out on his own. "Yeh," he vamps, "and as the structure gets bigger than a simple economy, it suggests something stronger than string. Maybe chains. Really big chains. Of course, chains imply slavery and undesirable bonds. They also make an uneven curve when you draw the ends together. It's not an even curve . . . I wonder what would happen if you threw a chain in outer space; what kind of shape would that make?"

I smile my trademark smile, disorganized, unevenly curved—not what you would call an integritous smile, but the smile of a man who has been outclassed. Guess I'll have to leave town.

July 19–25, 1995

7

ME MYSELF

A CODE

A Plea for Nonsense

GOSH, YOU MUTTER, shuffling in place, I feel so useless. What am I going to do with my life?

Well, first of all, be brave. Because as you read this, the multi-billion dollar Human Genome Project is busy mapping out every gene in the human body, and every time they find a new one, they file for a patent faster than you can say "nucleotide base pairs." In the process, our society is coming to resemble the movie *Brazil* more quickly than you might imagine.

Take the case of John Moore. According to Andrew Kimbrell, writing in *The Human Body Shop*, the University of California got patents on cells formerly belonging to Moore (that is, formerly belonging *in* Moore), and made a killing on several products derived from them, including Immune Interferon.

When Moore—who was in this mess in the first place because he had leukemia—tried to get remuneration according to the market value of his cells, the court ruled against him. You can't just hawk your body parts to the highest bidder, they said. On the other hand, the court also ruled that UC could continue to make a profit to its heart's content.

No surprises there. The stakes are high in the body scrimmage, and some of the players are pretty damn serious. But if UC is a big lummox with an attitude problem, then Craig Venter is a veritable linebacker on crystal meth. In 1992, while working for the National Institutes of Health, Venter filed a patent application on some 6000 gene fragments. That could turn out to be six percent of all human genes. Thanks to Venter, we have reached the moment when human genes can be discussed in *market shares.*

This news has not been received with hearty applause, even in Venter's own camp. Then again, morals are not necessarily the guiding lights here. Venter was able to file such a massive application largely because he has developed a high-speed "tagging" method that acts as a shorthand for an entire gene. In other words, he doesn't really know which genes he is patenting. As far as he is concerned, he can find that out later. Meanwhile, his competitors are left in the dust, plotting out genes the old-fashioned way—and they're not too happy about it.

To add insult to injury, even if Venter did know what genes they were, he might never know their functions. It's turning out that a lot of the genes have no known functions at all. For all we know, some parts of the genetic code are just a form of nonsense, a mindless prattling on the evolutionary line. The Patent Office, which can't give the green light to an invention unless it is useful, is thus eyeballing Venter's application with a healthy share of skepticism.

Now, you might think Venter is not so different from a conquistador standing on the shores of North America and declaring it the property of the Spanish crown. I happen to see it differently. To my mind, Venter is more like a second-rate musician trying to patent the notes C through E flat. If you're going to go after something so preposterous, why not reach for the whole thing? Why not patent every note in the scale? To drop the analogy, why work so hard to patent a mere *part* of a person? Why not go the proverbial distance and *patent yourself*? Wouldn't that be the safest protection in an age of genetic racketeering?

Very funny, you say. Too bad you can't really do it. Well, maybe yes, maybe no. I recently asked a former Patent Office employee if he thought a patent could be granted for a person. On the condition that I didn't use his name, he said it probably could. He wasn't too keen on telling me how to go about it, though, so I've had to navigate a possible path for myself.

As I see it, you would have to patent your genome, the sum of all your genes. But mapping it out, as the Human Genome Project is doing, will run you a tidy $3 billion. Venter's tagging method, for all its advantages, is also extremely expensive.

Not to worry, though—you can just get a DNA fingerprint. To do so, all you need is about $1000 and access to a standard biochemistry lab. The procedure is not so painful. A small piece of your DNA (obtained from, say, your spit) gets clipped into little pieces in such a way as to create a unique set of gene fragments.

Already you've overcome a bigger hurdle than you know. When you invent something, you usually have to conduct a search, which determines whether anyone else has arrived at your idea before you. When patenting yourself, however, you begin by knowing that your invention is unique, because it's you. The DNA fingerprinting procedure *is* the search. Just send it off to your lawyer and expect to pay the normal (read, exorbitant) filing fees.

But here's the payoff. The only objection to Venter's tagging technique that's being taken seriously is that he is trying to patent genes without knowing their functions. You, on the other hand, will run into no such trouble. When you patent your own personal genome, you will already know the function of every last doodle of DNA in your body, because *your genes are there to make you singular.* If the Patent Office tries to argue that you have no function, simply point to your Boy Scout merit badges, or your knack for peeling an orange in a single continuous strip.

No doubt some people will wonder why I'm giving this idea away. But truth be told, it's impossible to steal such an invention. Once you've patented yourself, nothing prevents your neighbor from doing the same, because your genomes are, by definition, different. It's a game everyone can play.

I also think it's fair to say that the ramifications could be far-reaching. First and foremost, the mere filing of your patent would change the nature of the debate. While Venter tries to patent genes on the basis of fragments, you try to protect your genes on the basis of your genome. Thus, in one fell swoop, the gene rush becomes organized not around the value of certain genes but around the value of individuals.

And you won't end up like John Moore, robbed of your ribosomes and penniless in purgatory. There will be sneaky Petes, for sure (there always are), who try to claim that your patent doesn't cover any disease you produce. You, however, will seize your long-awaited chance to play Shylock, and tell them they can take all the diseases they want but not a drop of the healthy stuff. At this point, they're sure to slink away.

Finally, with your patent safely in hand, you can sit back and relax. The feeling is uncanny, and oddly empowering. You're no longer at the mercy of the men in lab coats. And you don't feel useless anymore. No sir. You're the master of your own mysterious nonsense.

October 20–26, 1993

DNA Copyright as Performance

ALL RIGHT, I ADMIT IT, I've never had an original thought in my head. But at least I've got my original head, and the document to prove it.

When I last weighed in on the subject of genetics, I reviewed some of the midnight maneuvers taking place in the biotech world, and I offered a little primer on how you could patent yourself in a few, easy-to-follow steps. Well, what do you know but someone else thought of it first. Not that I'm upset—Larry Miller thought of a lot of things before I did. In his varied career as a Fluxus artist, he has undergone hypnosis in order to assume the identity of his mother, arranged bugs into football formations and masturbated onto a circular mirror.

Then in 1989, he decided he would do something kind of experimental.

We're sitting at the kitchen table in Miller's loft, hashing out the vagaries of genetic protection. In his wry, even tone, Miller is describing why he went to the top of the Empire State Building to copyright his genetic code. "I was thinking, you know, 'King Kong, Empire State Building.' It seemed like the magic of art or something. And all I had was a handwritten document."

Having declared himself his own creation, Miller soon became a kind of Kevorkian for the living and began to help others to do the same. He has had no trouble finding volunteers, either. To date, he has witnessed the copyright of about 400 people. At one point, he even tried to make his operation legally binding, but the lawyer who counseled him thought the matter should be handled as a patent, and this was beyond Miller's means. "I haven't had much luck on that front," he concedes.

Up until this moment, I had thought patents were the way to go, too. After all, patents for human genes already exist, and that's where the battle lines seem to lie. But Miller's commonlaw, para-para-legal strategy undermines the entire argument. By proceeding without government imprimatur, he is, in effect, questioning how it comes to be that an institution decides whether you or me or the guy who spits in a cup for a living can own himself or not.

"I've been accused of being a Luddite." Miller smirks flatly at the thought. "But I'm not against science or technology. All I want to know is, whose science is it?"

Most people don't think Miller is a Luddite, however. They think he's a comedian, which may be worse, because however much he enjoys the jocular veneer of his project, he is deadly serious about the stakes. "Here we are, on the verge of the genetic age," he says, "and so few people are aware of what it really means. It's a profound moment, because we're about to step into the business of evolution in a very direct way."

Indeed. If a biotech future begs for science fiction plots, there are certainly plenty at hand. Miller for his part, foresees a time when parents will have to subscribe to their children, paying the corporation on a layaway plan. My own personal premonition, which I may or may not have thought up myself, tells me we're destined to see the advent of *interactive genetics*, and after I bring forward a few salient facts, I'm sure you'll agree.

DNA, as you may know, consists of millions of base pairs denoted by the letters G, A, T and C. When a single base pair mutates—from, say, a G to a T—the results can be truly significant. Right now, genetic engineers are harnessing such mutations in order to white-out cystic fibrosis genes and the like.

So far, so good. Now imagine for a moment if many, many base pairs were to mutate. Our genes, as the biotechnicians are so fond of telling us, differ from the chimpanzee's only by very little. In fact, mammals are genetically more similar than different, so someday in the not-too-distant future, it should be possible to change humans into dogs.

Well, all right then. I say it's only a matter of time before these techniques lead to a new one—the all-American technology of *impulse mutation.* And who would be a more likely candidate to demonstrate the safety of this bold new endeavor than the President himself? When our able leader gives his usual landmark speech, the cinematic-industrial complex will simply supply each audience member with four pushbuttons that correspond to the base pairs of DNA. Then, as the genetic code of the President is scanned, the audience need only press a bit here, press a bit there, to mutate his genes by majority impulse. Think of the excitement the kids will feel as the Commander-in-Chief crosses species at their every whim. Hey, is that Tommy who's giving the President a donkey head? Or is Nancy doing it?

I tell you, this country isn't finished yet.

Okay. All this may be riveting enough, but while I've been brewing up nightmare-skits like this one, Miller has been exploring a few consequences out in the real world. One March, for example, he spontaneously offered to buy the genetic code of one Willem de Ridder, a German artist. Much to his surprise, de Ridder accepted. (The sum, by the way, was 10,000 Korean won, or about $10.) Since then, two others have sold their codes to Miller, leaving him to wonder just what he should do with these unusual properties. Should he try to put his newly purchased genetic codes on the stock market?

"At some point," he explains, "I do want to transfer somebody's genetic record to another party. I probably will ask the person first, even though I don't have to. I would even like to see a court case on something like this."

But now we have reached the moment of truth, and over we go to the far end of Miller's loft. I proceed to fill out my own Genetic Code Copyright certificate: mother's name, father's name, place and date of birth. A fingerprint and a signature later, Miller takes my hand and places it squarely on the NYNEX White Pages.

"Do you swear that you are an original human?"

"Yes," I intone.

"Never been abducted by a UFO or been in circumstances where you could have been genetically altered?"

"Not to my knowledge."

Miller looks me straight in the eye and says he believes me. He signs as a witness, then impresses a fingerprint of his own. "Congratulations. You are now a certified original human."

The top of the Empire State Building it was not, but I must admit it felt pretty good to become unequivocally, officially original. As for what I'll do next, I'm not exactly sure. Maybe I'll hang the certificate on my office wall. (As Miller points out, "Sometimes people think you're a doctor.") Then again, may be I'll try to file it down in D.C., just to see what happens.

December 8 – 14, 1993

Dinner with a Geneticist

LATELY, EVERYONE and his brother has become curious about my attempt to copyright myself. The letters are jamming the transom; the faxes are rising in piles. When, their authors inquire, can the plaudits from the scientific community be expected, what will I do when my genome is rightfully mine and just who the hell do I think I am exactly, anyway? Why, just the other day, a woman wrote me a letter proposing marriage, on the condition that she be allowed to purchase my rights.

As with any event of this magnitude, the particulars have been all but completely obscured. And so, reluctantly—*phlegmatically*—I have determined to rouse myself from the reading room, extinguish my meerschaum before its time and once again take to the metaphorical lectern.

I can think of no better way of clarifying my enterprise than to recount the events at a dinner engagement I attended recently at the establishment known as El Teddy's. I had gone there to meet a friend, but when I arrived, I found that my friend had been joined by his cousin—one Jonathan Jones, director of the genetics laboratory at the Sainsbury Institute in England. I immediately remarked the fortuitousness of our meeting, as did he, for we had heard of each other. What's more, Mr. Jones is an Englishman with firmly held beliefs, and as I am descended from a British bastard somewhere along the line, it was destined that a kinship form between us. We fell at once, as they say, into earnest social intercourse.

Mr. Jones, it should be noted, has worked in the field of genetic engineering for many years. In the

1980s, while teaching at a biotechnology company in California, he was engaged in inventing a protectant for crops that would freeze at one degree higher than common water. The machine point was this: when sprayed on crops, the protectant would form a shield against an oncoming frost. Although this endeavor ultimately failed, he and his men performed famously, sacrificing long hours to the hope of its success without thought for their own gain in the slightest.

Sometime thereafter, Mr. Jones returned to Sainsbury and undertook a genetic engineering experiment in an effort to render the ordinary potato resistant to the ravages of the potato blight. As you may recall, it was the potato blight that caused so much suffering among the 19th-century Irish, while the British looked on in admiration at the quaint Celtic capacity for survival, no matter how grueling the circumstances.

In the event, we had heard that Mr. Jones had been making good progress in his experiments, and indeed, he had come to America with the express purpose of communicating his latest advances to his colleagues at New York University. I commended him wholeheartedly on his promising work, with the added encomium that, even with the tremendous pressures under which he had labored, he was more than a century late.

Mr. Jones must have gathered my tender sentiment entirely, as he fell silent for a moment, his face deepening into a blush. Composing himself at last, however, he managed to inquire as to my experience in the field of genetics.

I spared no expense in informing him that I had applied for a copyright of my genetic code under the category of a Musical Work. "Each of the four base pairs that make up the celebrated double helix has its own identifying frequency," I explained, "and as such, must vibrate at a unique musical pitch. Because the combination of these base pairs in each individual is unique, each individual human being is composed of a unique set of pitches. Therefore, a human being can legally be categorized as a musical composition."

In addition, I continued, losing fractions of my ceviche to the air, the Register of Copyrights requires a "material deposit." In the category of music, a "phonorecord" is considered acceptable. This was a felicitous fact in my case, since I could simply deposit a fingernail in a small plastic envelope and send it off to Washington along with my application. Inside the fingernail was contained the entire musical work of myself—expensive to play back, there was no doubt, but contained in the item nonetheless.

By now, we were fast friends; as I could ascertain by the manner in which he coughed—a repeated hack during which several portions of burrito regained an altitude previous to their swallowing. "That . . . is the most . . . preposterous thing I've ever heard," he sputtered. "Where is the work? Who has done any work?"

Naturally, I engaged this sally with my best defense. Drawing on an innate theatrical bent, I wove a heartrending tale of my mother, who had endured long years of suffering, not only in ensuring my birth but also in furnishing my genetic code with every form of protection imaginable. As you clothed the crops with your protectant, I lowed, so did my mother clothe me, and it was she who was the author, with some help from dear dad, of my particular minuet of DNA.

Yet perhaps I had worked my passion to tatters, for I saw Mr. Jones drifting into a feeble state of agreement. In order not to lose him completely, I pressed my point home: here in America, I said, the 13th Amendment of the Constitution prohibits the ownership of one person by another. Therefore, I had contrived to apply for my copyright, not as the author, but as what is called the "copyright claimant." Even a moron, I hinted, would see the sense in this.

"And how do you expect to make money off your genetic code?" he rejoined, now happily back in the fray.

I felt the mistake as soon as the words had passed my lips. "I don't expect to make any money," I replied, "unless there is money to be made."

"Oh, I'm glad to hear that!" he cackled, dancing figuratively on my grave.

At this point, the majority of our discourse had been accomplished, and we passed the remainder of the meal in that variety of mutual humiliation that is noted for its squeamish pauses. On the odd occasion, he stole several portions of my ceviche, all the while affecting the downcast gaze of a criminal. This habit caused me to deride him tactfully, until it was revealed that the fare before me was in fact the seafood salad he had ordered. Thus was the meeting of our minds complete. As I surrendered the salad over to him, he remarked with quiet resignation, "I suppose you could say that it throws the issues into sharp relief."

For this I was only too happy—for what is the aim of genteel conversation if not to induce an unstated shame? Indeed, my euphoria might have continued much longer had I not heard reports of Mr. Jones' lecture at the university the following afternoon. In recounting the epic of the Great Potato Famine, he was alleged to have laid the blame on the potato blight itself. Stopping himself in mid-sentence, however, he ejaculated: "Well, actually, we British had something to do with that."

I cannot express the disappointment I felt at hearing this sudden outburst of guilt. Some things, after all, are best kept between gentlemen.

December 28 – January 3, 1995

Getting Permission from Mom

THE REJECTION LETTER from the Copyright Office sat in my drawer, gathering the pale municipal dust that once was my desk. Should I leave it at that? I wondered. Should I relegate this odd gesture to obscurity, to be stumbled upon in some distant age when our alarm-packed century has become harmless and quaint? I turned the matter this way and that, letting it reflect the mysteries of phenotypes and family trees, until at last I knew what I had to do. It might not be easy, it might not work, but it was time to call Mom.

"No, no." She roused herself from a half-drowsy trance. "No, I was just reading."

We exchanged only the vaguest of greetings before I jumped in. "Mom, you remember how I said I was trying to copyright myself?"

"Yeh, I think so," she yawned. "You—you wrote about it, didn't you?"

"Yeh," I answered, slipping into my stride. "I also sent a form to the Register of Copyrights, trying to do it for real."

"Yes, I think I remember that."

"Well, now they've sent me a rejection letter, and I'll tell you, when I read it, I was tickled silly. It's much better than a form letter. It gives a whole list of reasons for rejecting the copyright of my genetic code as a musical composition, and there's even a name signed at the bottom. That means I can write back to someone in particular." I hesitated, my voice off balance. "Which brings me to the reason I called."

She offered a tentative consonant or two. I could hear the TV rambling in the background.

"You see, the way I wrote it," I continued, "I didn't claim that *I* was the author of my genetic code. I said that you and Dad were, and that I was the copyright *claimant* under the 13th Amendment, which is the anti-slavery amendment."

"Yehhh...?"

"Yeh, no, anyway, listen to the letter. It says, 'If there were a musical work of human authorship other than your own, you would have to have obtained ownership of any copyright therein through a written instrument or by operation of law.' So you see, that means, ummm—"

My mother may be a lot of things, but slow on the uptake she is not. "That means," she deduced, "that you could be my slave."

The thunder of eons rattled my bones. "Well, I guess so, yeh."

"Interesting." She let out a short whinny of a laugh. She was definitely awake now.

"Anyway," I said, hoping to breeze past the unseemly business of servitude, "this guy is essentially saying that I need some kind of written statement showing the transfer of ownership to me. So you'd have to be willing to call yourself the author, or the co-author."

"What does that mean actually? That you're my son?"

"Well, it—"

"I'm willing to say that you're my son."

I think I blushed at this, even if it carried a whiff of a mother defending her son at an arraignment. "Thanks, but actually, it's a little different from that. For example, the Copyright Office could argue that authorship requires a certain amount of conscious effort. Which, really, when you work it all out, means that I'm asking you"—pausing before the punch here, I took a deep breath—"*whether you planned to have me.*"

Now we were both embarrassed, and her voice softened. "Well, what do you think?"

I thought about my parents. I was forced to admit that I didn't know either of them all that well before I was born.

"Yes, Dave," she said at last, "we decided to have you."

So far I thought I was doing pretty well. I had brought her in as a player, then distracted her with sentimental pap, getting my answers as I went. From here I figured it was pretty much of a cake-walk. But it was not to be, not just yet. Having agreed to the title of My Authorship, my mother moved seamlessly into the role of interrogator.

"So tell me—this is just a joke, right?"

"Sort of," I said. Then I decided to pull out all the stops. "But it's also a kind of thought experiment, because nobody really knows what will be possible with genetic engineering in ten or twenty years." I briefly described how DNA had already been used as a computer. I was going to add that the Department of Commerce recently took genetic samples from a Central American woman under dubious circumstances, and then follow it up with a swift excursion through cloning, but she stopped me before I could make much headway.

"What's that? DNA acts like a computer?"

"Oh yeh. It was in the *New York Times.* A guy in California used bits of DNA to figure out the traveling-salesman problem."

This problem, I explained, involves calculating the shortest distance between seven cities. Humans generally can't figure it out without trying every combination, and if you start adding more cities, even a computer will grunt under the strain. It's been a real bugger—until recently, that is. Last year, a certain Dr. Leonard Adleman at the University of Southern California got fragments of DNA to solve the riddle in no time flat.

"I don't understand that at all," she said, a hint of exasperation entering her voice.

Had I been able to whip out the *Times* article of November 22, 1994, I might have exasperated her even further. I might have ventured that Adleman made gene fragments attach themselves to one another with amazing economy, leaving no doubt that they had solved the problem. This would have left my mother confused, but at least it would have approximated the idea. As it was, the article lay buried beneath several inches of fax paper and notes, so I could only offer a more ponderous riff involving alphabets and the Enigma machine and, I think, jelly.

"I really don't need to know how DNA can be a computer," she said. "But tell me: if you copyright your genetic code, what will it do to your children?"

Ah, the grandchildren clause. Always a sticky question, even in the most conventional of mother-son conversations. But I was banging on all cylinders now. Leaving aside for the moment the conspicuous absence of offspring in my life, I suggested that I would copyright their pink little codes fresh out of the womb. Then, when the day came, I would sit them down, pat their furry heads and sign away my claim to their genetic codes as I regaled them with tales of adulthood.

Miraculously enough, that seemed to resolve it for her. "Well, just tell me what you want me to do."

"I guess I'll write up some kind of statement for you to sign," I replied. "And we should probably go to a notary public. I mean, they'll probably think I'm crazy, but it's worth a shot."

"Maybe you are crazy," my mother suggested. "But I'm willing to humor you."

Humor me? Did my mother say she would *humor* me? Ah, the mysteries of phenotypes and family trees. I had come as close to unconditional love as a son can possibly get.

March 22–28, 1995

The DNA Copyright Project

THE 1996 COMEDY *Multiplicity*, in which Michael Keaton plays himself and several clones, should be proof enough that genetic engineering is headed for disaster, yet the industry proceeds regardless. And so do I.

As faithful readers of *New York Press* are doubtless aware, for the past few years, I've been slouching toward Bethlehem with the rude idea of copyrighting my own genetic code. From the start, it only seemed logical that my own biology should belong to me. In fact, I'd be much harder pressed to justify the ownership of anything else. Others have seen it differently, of course, and they've done their best to nip the project in the bud. But they've only succeeded in spurring me on to a larger campaign. What doesn't kill me makes me a much bigger pain in the ass.

This being a column devoted to inventions, I initially framed my perverse little thought experiment in terms of patent law. Since genetic engineers were already patenting genes, even human genes, the U.S. Patent Office seemed a natural target. Then I met up with Larry Miller—a Fluxus artist who was copyrighting people's DNA in an unofficial, ritualistic way—and became convinced that copyrighting made more sense. Not least compelling was the cash factor; a patent can run into four or five digits, while a copyright costs only $20. Even better, DNA has to be *copied*—until there's a whole jamboree of the stuff—in order to be useful. The elegance was too much to resist. I switched over to copyrights, with the intent to apply for one for real.

There were plenty of parameters to work out, some easy, some less so. For starters, I couldn't honestly claim to be the author of my own genetic code—that honor belonged to my parents. No problem there: I simply applied as the copyright *claimant*, arguing that my parents no longer had any rights to my code, thanks to the anti-slavery imperatives of the 13th Amendment.

Fitting into the predetermined esthetic boundaries proved to be a tougher proposition. Forced by the protocols of law to choose a category, I decided—after some ridiculous soul-searching—to register my genetic code as a musical composition. DNA, I reasoned, is made up of four chemicals—guanine, cytosine, adenine and thymine—that appear in roughly the same sequence throughout the human species, except for the slight variations that make each individual unique. And what is every sonata and soul tune ever written, if not the same 12 notes worked and reworked in endless variety?

But what clinched the musical analogy was a loophole in the law. For musical compositions, the Register of Copyrights allows "phonorecords" to be sent as a deposit and as it turns out, phonorecords are defined as "objects embodying fixations of sounds." Well, that settled it. Everything in the universe vibrates. Everything that vibrates creates a sound. Bingo: one thin shard of fingernail embodies the whole of my grand genetic opus. Pleased as punch with this realization, I dropped a dry white crescent into a tiny baggie, sealed the envelope and slipped it into the corner mailbox.

A good six months elapsed before I received a response. Naturally, the examiner, one James Vassar, served up his rejection with great ladles of bile. That much I expected. But he also managed to go too far. Rather than dismissing my claims out of hand, he felt compelled to offer detailed reasons for the rejection. And earnest student that I am, his articulations allowed me to hone my arguments and proceed to the next bloody round.

For example: Vassar wrote, "If there were a musical work of authorship other than your own, you would have to have obtained ownership of any copyright therein through a written instrument or by operation of law." The 13th Amendment, he added, "has nothing to do with transfer of copyright ownership."

I took this point to heart and did the only thing I could: I called my mother. After an odd, circuitous discussion, during which she noted with some pleasure that I was "her slave," she agreed to sign her ownership of my genetic code over to me in the presence of a notary. Off we went to the shopping center in downtown Ridgewood, New Jersey, and delivered our documents to a testy gentleman who was simultaneously running (or trying to run) a dry-cleaning business. I doubt he had any idea what he was stamping, even though the title across the top blared out **TRANSFER OF GENETIC CODE** in eye-popping caps.

One of the examiner's caveats was dead, at a cost of $2. But there were more caveats to go. Vassar also had reservations about the originality of a genetic code. "Even assuming that normally inert chemical substances can be excited into vibration," he wrote, "any resulting acoustical phenomena do not meet the definition of a copyrightable musical work because they would not be original and creative products of human authorship."

What Vassar was saying, in so many words, was that screwing is a mindless act undeserving of the title of creative work. This in itself was a fairly strong point, as anyone who has ever had sex can attest, but then he went on to cite copyright law: "For a work to be copyrightable, it must owe its origin to a human being. Materials produced . . . by nature . . . are not copyrightable [Op. Cit. 202.02(b)]."

Really! To what does a human being owe its origins, if not another human being? I mean, anyone will see the resemblance.

Still, Vassar pulled a worrisome exegesis out of this hat. "Implicit in the concept of 'writings'" he rifled, "is that it represents conscious, intentional expressions of mind." While that judgment struck me as fairly cavalier (to say my mother was unconscious at the time—the idea!), Vassar was undoubtedly the judge and I the subject in need of a good comeback. So I got in touch with someone who knew a thing or two about "intentional" writings.

Those acquainted with William Burroughs from his pre-Nike phase will probably remember the joys of his "cut-up technique." Essentially, he took a page of manuscript—on which, more often than not, he had detailed some horrible atrocity—and cut it in half. Then he did the same with another piece of paper, and another, until he could shuffle and match the scraps to create surprise sentences.

Clearly, the result of this activity was not the "conscious, intentional expressions of the mind." Or, to put it in Burroughs-speak, it *was* an expression of the mind—but whose? His collaboration with Brion Gysin gave rise to a book that contained the very nut of the mystery in its title: *The Third Mind.* Nevertheless, according to his assistant, James Grauerholz, Burroughs has always sailed straight through the copyright process. "As for your question about copyrighting examples of 'cut-ups,'" Grauerholz wrote back, "no, we have never had any problem with that; actually, as both William and Brion have written, cut-ups are not really 'random,' both because there is authorial selection, and because (as they put it): 'How random *is* random?'"

The parallel, I trust, is obvious. Did my parents not exercise authorial selection when they took their pre-wedding blood test? And did they not produce a work in me that was just as random or organized as, say, *Nova Express*—or the music John Cage composed with aid of the *I Ching*—when they made the beast with two backs?

You have to admit, the argument is gathering steam. In fact, all that remains is to get a scientist, preferably a highly decorated one, to verify the vibrational characteristics of the four chemicals present in DNA. Once I've got that, anyone with $20 and a dream will at least have a case for copyrighting his own genetic code.

Imagine 200 people—or 500, or more—sending in their applications *en masse.* At best, it would change the nature of the biotechnology debate forever. At worst, it would just be a gas, and a far better premise for a movie than the usual cloning shtick. Any takers?

July 31 – August 6, 1996

Coda

WHENEVER THE WORLD becomes too much for me to handle, I like to think about Tibetan sand paintings—not just the fact of them, but the whole system by which they're made. Which is to say, the whole system by which they're *un*-made. Tibetan Buddhists, as a rule, do not approach a painting in the manner of a Rauschenberg or a Rothko. They work at it, they sweat at it, they spare no expense to limn heaven or hell, demons or angels, what have you. Then, when the painting is done, they take it outside and celebrate while the wind blows all of their efforts to naught. Only the counterfeit article can be found in a museum; the real ones are always temporary, always defiantly mortal.

That kind of metaphysical heave-ho, with its attending release, seems especially relevant this week as I celebrate the literary destruction of "The Patent Files." And in a case of double relevance, it just so happens that I recently met a man with exactly the right method for dispatching my column—and the illusions it came in with—back to the hoary elements.

The man can be described quickly enough. Paul Roossin worked for seven years as a neurobiologist before taking on a 12-year stint at IBM's Watson Lab, where he helped develop MedSpeak, a large-vocabulary, continuous-speech recognizer for radiologists. It's not hard to imagine that he took to R&D work right away. Bright-eyed, energetic and possessed of a quicksilver mind, he fits the mold of the boyish technologist with ease.

The method, on the other hand, requires some explanation. More specifically, it requires a passing familiarity with a Borgesian item known as a Markov chain.

"Markov," explains Roossin, scribbling in my notebook as he talks, "was concerned with transitions from one state to another. But the states themselves weren't important. What he was looking at were the paths between states. Not only that, but he looked at the situations where more than one path could be taken and developed a statistical framework for analyzing what happened."

What this boils down to is a system that has the same basic outlines as hypertext. Imagine a hypertext program that moves through branching pathways of windows of its own accord, using statistics to make its choices, and you'll have a basic idea how a Markov chain works. Then, instead of windows, imagine that the units are words: the program comes across a word, scans the text for every other place where that word appears, checks all the pathways from that word to the next and chooses a second word accordingly. Then it looks at the new word and repeats the process, and so on, until it hits a randomly chosen word, at which point the whole shebang comes to an unceremonious halt.

Make no mistake about it: a Markov chain is no respecter of syntax. It has nothing to do with artificial intelligence or the imminent takeover of the alphabet by heartless machines. This is painfully obvious when you're dealing with the simple word-to-word pathways called *bigrams.* Nevertheless, a Markov chain will begin to deliver a *facsimile* of human thought when it goes after trigrams—that is, as it begins to bundle two words together in search of a third. And by the time you get up to quadrigrams, which look for a fourth word to follow a found example of three, the machine is reading out chunks of text verbatim.

Curious to see what Markov would make of my handiwork, I asked Roossin to feed 10 different "Patent Files" of years past into his computer. What it spit out in quadrigram mode came as something of a shock:

Well, this is the point where my expository skills begin to careen out of control—and rightly so, because by the time I've had my customary bimonthly thought, the sheep have already outrun me. Still, I feel reasonably sound in body and mind to venture a provisional reading. First of all, I think we have to admit that we don't rightly know what a human is anymore.

I wrote that passage late one night when I thought no one was looking, quite unaware that some truculent non-human already had me pegged. Nor was this an isolated case. The better part of this 8000-word text could have passed for a rough draft of my own writing. Even when a quadrigram chain did veer into non sequitur, it did so with a certain amount of stealth:

From there, it's a matter of dragging your finger to the chapter you want. Lift your finger and the first page of the chapter materializes. Touch the outside corners of the room. Nothing in those corners will give him the answer, yet off his eyes go, this way and that. If the brain is such a beleaguered master, it has to be able to use this model, come up with some values for it and put it out on a computer spreadsheet. The strings are the exchanges, the rods are suspended in mid-air. The strain on every segment of string is exactly equal.

That still passes for a facsimile of human thought, even if the author is no longer in complete command of his subject. Such pretenses to sense begin to fade upon arriving at the level of the trigram. Still, I could almost hear the plagiarist at work, wheedling elegy from my exposition:

Most bright young ideas never see the latest deus ex machina. As it was, it was just a nice guy from Houston whose brainwaves travel in a hotel . . .

A garbage truck grumbled by. Two women in furs were swept down the sidewalk on the air, and when he learned what the Cubans had in mind for their lassoes, "What would my wife think if she saw a . . . card in my wallet?" asked one man, who presumably had something going on here with anti-cloning."

And what am I to conclude from this passage except that the universe considers my disordered thoughts a ripe vein for the pickax?

I'm duty-bound to applaud anything that can be seen from almost any angle. Certainly, a good deal of processing takes place by mistake. Say a duck's genes go a little . . . off, giving its offspring longer bills. If these longer bills prove useful to survival, then the longer-billed ducks will start to outnumber their shorter-billed peers until the long billers are the people who don't answer the phones. It's not necessarily the gravest scenario. Others imagined a robber baron descending from the darkness.

Oh yeah. I seem to recall Darwin saying the same thing.

Of course, this game can be played *ad infinitum*, to the limits of one's willingness to have one's brain rearranged, but there's only so much time in the world, and the fact of the matter, as I've already said, is that I have a mission to fulfill. So . . . is there any time left for the final bon mot before progressing to the bigrams and their inane Caliban prattlings? Any seconds to spare on the ultimate meaning of man and machine and all the vicissitudes between? Any last chance to prize free that last, rare grain of truth before the sand is swept off by the winds, never to return?

Is there ever?

What a dissident political terrain known as a professed advocate of the electronic book to know from the book to David Lindsay. One is written on the food processor, he has had no such gestures as a genuine event . . . Simpleheaded as he was, you can think if every few others, but lately he's also set up on the card and obtained a page. They have seen. Open the clear sun of the Greek island where telemarketers make an appearance around a mast, and you've got the clear sun of defense and dad chose to make matters much worse . . . And will be called sex in your brainchild a series of the '70s and what deserves to sleep at lottery-ticket outlets, no responsibility for had the landlord in America. I must have to provide one of invention made by anyone had moved back to make it home, and I to make periodic calls a meat grinder . . . To

Washington, compadres, your brainchild a catty-cornering daydream: What can be the medium like an optimistic past 25 years. Of course, call a phonographic device that wire, because they would eventually come up for every U.S. laws? Somehow I disagree. Or I'll stop imagining things like many sexually transmitted diseases, to David Lindsay **THE PATENT FILES** *Percentage: I'd like car alarms wailing unholy dissent. "The Patent Files" takes no responsibility for the chain after that . . .*

March 25 – 31, 1998